FOOD ADDITIVES

Characteristics, Detection and Estimation

OTHER BOOKS
(Published by APH Publishing Corporation)

- Food Contaminants—Origin, Propagation and Analysis
- Food Safety—Concept and Reality
- Potable Water—Indian Scenario
- Food Irradiation—Indian Scenario
- Food and Nutrion Education
- Beekeeping

FOOD ADDITIVES

Characteristics, Detection and Estimation

S N Mahindru

Emeritus Consultant
Food Standards and Laws
MDH Ltd., New Delhi

A P H PUBLISHING CORPORATION
4435-36/7, ANSARI ROAD, DARYA GANJ
NEW DELHI-110 002

Published by
S.B. Nangia
A P H Publishing Corporation
4435-36/7, Ansari Road, Darya Ganj
New Delhi-110002
☎ 23274050
Email : aphbooks@vsnl.net

Rs. 3995/-

2026

Typesetting at
Paragon Computers

Printed at
Balaji Offset
Navin Shahdara, Delhi-32

Foreword

The manufacture and marketing of traditional as well as non-traditional foods is an important activity in our country. There is a time lag between the preparation and final consumption of these foods necessitating their preservation in the original fresh state. This becomes possible with the use of food additives such as preservatives which protect the food from microbial decay, anti-oxidants, anticaking agents, emulsifying and stabilizing agents, flavouring agents, sequestering agents and buffering agents, etc. thereby improving the longevity of foods. All these additives are chemical in nature and their addition is governed both in type and quality under different control orders of the Government.

The area of food additives has gained special importance during the liberalisation of Indian economy resulting in mass production of non-traditional as well as traditional foods. The regulation of their production, packaging, storage and transport also assumes great importance. For the manufacturing technologist, it is imperative to know the chemistry, scope of utility and analytical techniques of such food additives.

This book on *Food Additives-Characteristics, Detection and Estimation,* embodying relevant information culled out from the latest international and national literature on the subject is an excellent endeavour by Shri S N Mahindru, retired Chief Chemist from the AGMARK organisation and presently a Consultant Emeritus in Quality Assurance, Food Standards and Food Laws. There has been a long felt need for a consolidated document on this important topic specially in the Indian context. This publication is very timely to fulfil the gap. The subjects of low calorie sweeteners and flavouring agents which have gained special importance in the present day life have been dealt in greater detail with unconventional information.

I consider this compilation a valuable reference guide to the food processors and analysts. It would also prove useful to those who are engaged in enforcement of food safety laws.

R A Mashelkar
Ex. Director General, CSIR
and Secretary, Department of
Scientific & Industrial Research

Preface

Food production is on the increase worldwide. Various technologies have enabled the availability of these foods, and the products derived from them, over long distances and round the year by prolonging their shelf life through the application of food additives.

In order to regulate the production, manufacture, packing and packaging, storage and transport of these foods so that the health of the consumer is protected, the Government of India keeps a vigil through the punitive law known as the Prevention of Food Adulteration Act, 1954 and the Rules framed thereunder. The Act and its Rules in the present scenario cover the standards for various foods, as also the use of food additives in addition to other steps for controlling food adulteration and wholesomeness.

The subject of food additives has gained importance after the recent liberalisation of the Indian economy resulting in mass production of occidental foods mainly for export. These foods cannot be produced without food additives. This book has been prepared with a view to acquaint the Indian producers, the quality enforcement personnel as well as the judiciary with the rules regarding the extent, and the method of use of these additives.

The analytical methods described in the book are followed in the Government laboratories only. I have learnt and practised these methods during my tenure in the Agmark Organisation of the Government of India from 1954 to 1988. During my frequent appearances as a defense expert in P.F.A. Act cases in various courts, I became aware of the lack of information with respect to the use of various food additives in the food industry, the quality control personnel and the advocates specifically defending clients in the P.F.A. Act cases. This book aims to acquaint the above personnel with the precise laws related to food additives so that involuntary infringement of analytical procedures can be prevented.

I have used my vast experience and in-depth knowledge base to describe the precise methods of analysis practised in the statutory laboratories. Many of the topics though inscribed in the statute are little known and practised, viz. analysis of flavours, vitamins and emulsificants. These have been presented to make the average analyst comprehend them and perform the tests confidently and with reproducibility.

This **book** shall also be useful for candidates appearing for the Public Analysts Qualification Examination under the Directorate General of Health Services of the Government of India as well as for those appearing for the A.I.C. Examination conducted by Institute of Chemists, India. The book shall also be useful for advocates specifically dealing with Food Adulteration cases.

Towards the preparation and compilation of this book, I am indeed indebted to my several friends who are in this field, especially to Dr S K Saxena, Ex. Director, Food Research Analytical Centre (FRAC) of the Confederation of Indian Food Trade & Industry (CIFTI) of the Federation of Indian Chamber of Commerce and Industry and M M Madan, who is a well-known Food Processing Technologists. I am thankful to S/Shri Dharam Pal, Managing Director and Rajeev Gulati, Director, Mahashian Di Hatti Ltd., for their encouragement in several ways. I am grateful to Dr. R A Mashelkar for writing the foreword to this book.

My special thanks to Sudershan Khanna of Navyug Commercial College for computerisation of the manuscript in a short period.

S N Mahindru

Contents

Chapter 1

Food Additives: Chemistry and Estimation

INTRODUCTION

The food additives are used to preserve the quality of the food and maintain its appeal and at times to restore diminished nutritional value. Wastage of food is a national loss and should be avoided at all costs.

According to Prof. Sir Frank Young, food deterioration results when the food supply is more than the consumption. Deterioration can also occur during the long time required in transit or during storage before it actually reaches the consumers. The addition of permitted chemical substances, i.e. food additives can prevent or greatly impede the process of deterioration.

Alstair Frazer, a world authority on food additives, further adds that these additives help to achieve some degree of uniformity needed for large scale production, to enhance the flavour or improve the appearance or texture of a product and make it more acceptable to the consumer.

Principles governing the use of a food additive as summed up by Frazer are:

(i) Technologically effective,

(ii) Safe in use,

(iii) Absolutely necessary quantity,
(iv) Never with the intention of misleading the consumer about the nature or quality of a food and
(v) Minimum use of non-nutrient food additives.

NATURAL vs SYNTHETIC

There is a widespread belief that natural substances are safer than synthetic substances. Nothing could be farther from truth. Most of the potent known poisons occur naturally in plants, animals or microorganisms. Many natural substances of plant origin are complex and contain molecules that cannot be easily dealt within the animal body. Many constituents of foods are unknown and considerable variation may occur in the composition due to genetic or environmental factors. Synthetic substances are often simple and of greater purity.

Food processing has gained importance in the national scheme of economic growth. To regulate production of processed foods and safeguard the health of the people it is imperative to prevent production of sub-standard food products. This is being accomplished by voluntary quality ensuring systems like Agmarking of agricultural and live stock products, and I.S.I., certification of other food products. Compulsory quality enforcements are being exercised through orders like Fruit Products Order (FPO), Meat Food Products Order (MFPO), etc. Above all these is the Prevention of Food Adulteration (PFA) Act of 1954 and its Rules of 1955, amended and updated continuously through the Central Committee of Food Standards (CCFS).

Food additives under PFA Rules as mentioned under Part XIII A, Rule 64-C, include colours, flavouring agents, antioxidants, anticaking agents, emulsifying and stabilizing agents and preservatives. Only these have been included in this compendium. The presence of insecticides and pesticides have been excluded though these are initially employed to safeguard primary foods from the attack of insects or pests but are undesirable in the processed foods. These have been dealt in Mahindru's "Fond Contaminants—Origin, Propagation Analysis" pub. by APH Publishing Corporation, Ansari Road, New Delhi 110002.

Chapter 2

Preservatives

INTRODUCTION

Under Rule 52 of the PFA Rules 1955, a preservative means a substance which when added to food, is capable of inhibiting, retarding or arresting the process of fermentation, acidification or other decomposition of food.

According to this rule, preservatives have been classified into Class I and Class II preservatives.

Class I preservatives belong to the following four distinct taste groups,

(i) Saltish—common salt.

(ii) Sweet-sucrose, dextrose, glucose or its syrup and honey,

(iii) Pungent—spices,, vinegar (natural as well as synthetic)

(iv) Oily-only edible vegetable oils. Other fats like milk fat, animal fat or marine oils have been excluded.

Class II preservatives are,

(i) Benzoic acid including salts thereof,

(ii) Sulphurous acid including salts thereof,

(iii) Nitrates/nitrites of Na/K,

(iv) Sorbic acid including its Na/K/Ca salts,

(v) Propionic acid including its Na/Ca salts and its esters,

(vi) Lactic acid including its Na/K/Ca salts and Na diacetate,

(vii) CH_3 or C_3H_7 esters of p-OH benzoic acid,

(viii) Acid Ca-phosphate and

(ix) Nicin.

All these are chemicals and correspond to US Food and Drug Administration's definition of preservatives. According to FDA's definition, preservatives include antioxidants to prevent oxidation of oils and fats, pigments and flavours, anti-browning compounds irrespective of these being enzymatic or non-enzymatic, antistaling compounds to stop textural changes in foods such as use of emulsifiers in bread and antimicrobial additives to prevent or arrest the growth of microorganisms in foods.

In general usage, a food preservative is a substance which is capable of arresting or preventing microbiological spoilage of food products during handling, processing and storage. Besides, its daily consumption should not harm the human body over long periods. For example, formalin is a good preservative for milk yet is prohibited as it has toxic effects on health.

The following factors influence the selection of preservatives:

Properties of food

The suitability of an antimicrobial additive in food products is determined by its pH, composition, nature of spoilage normally encountered during processing, handling and storage, the physical and chemical properties of the antimicrobial compound, its sensory properties, safety as well as method of application. Loss of antimicrobial activity can occur if the preservative reacts with food constituents leading to binding of the chemical structure of the preservative. The pH of the food may ionise an active group in the preservative molecule and thus change its activity.

Some preservatives may be oxidised or hydrolysed and their sensory and microbial activity may get altered. Antimicrobial activity of sulphites and bisulphites is known to decrease considerably by their reaction with aldehydes and reducing sugars because of their having the terminal aldehyde group. Protein and fibre can also bind several preservatives resulting in a retarded effect. Some of the preservatives which are hydrophobic in nature tend to partition in lipid phase of food and get concentrated far off from water phase where the microbial growth actually takes place.

Level and type of microorganisms

The nature of food production prior to incorporation of a preservative can directly influence the microbial flora an.. tot..

microbial load. The method of food preparation also may influence the choice of the preservative to be used.

A preservative can never be a replacement for sanitation and hygiene in a food processing operation. Every endeavour must be made for a low microbial load to avoid a heavy concentration/dose of the preservative. As the antimicrobial spectra is different for each preservative, the nature of microbial flora contaminating the food product significantly influences the choice of the preservative to be used.

Effect of other preservation treatments

Antimicrobial preservatives are rarely used alone. Other processing treatments like pasteurisation, partial dehydration, refrigeration and packings are also required. The type of preservation method used in conjunction with antimicrobial preservative will also have a significant influence on the type and level of antimicrobial preservatives to be used. For example, lowering of water activity will accelerate the growth of osmophillic yeasts and molds, thus requiring the antimicrobial preservative capable of limiting the growth of these organisms. Refrigeration is generally selected for psychotropic gram-negative micro organisms. As packing alters the environment of the food it influences the growth pattern and type of micro organisms. Vacuum and inert gas packaging normally employed for protection against rancidity development invariably discourages growth of molds and several acerobic bacteria, but encourages the growth of certain anaerobic organisms.

Safety It is directly related to toxicology principles. For instance the quality of the substances ingested and absorbed is important regarding their effects on the body. It is possible to define a dosage level of a substance that is likely to have no demonstrable effect; greater amounts might have well defined effects; finally a dose will be reached above which no enhanced effect can be elicited. This dose/response relationship of an additive are of fundamental importance in the field of food additive safety evaluation. The food additive is commonly used in the no-effect dosage level, with a provision for an adequate margin. This no-effect dosage is physiologically harmless. However, good a preservative is in its antimicrobial activity and sensory quality, its use in foods cannot be permitted until and unless its safety is proved by animal feeding studies. Based on long term studies, World Health Organisation has laid down the acceptable daily intake (ADI) for various additives as given in Table 2.1.

Table 2.1

Preservative	*ADI (mg/kg body wt/day)*
Acetic acid including its Na/K salts	no limit
Sodium diacetate	0-15
Benzoic acid including its Na/K salts	0-5
Formic acid	0-3
Hexamethylene tetramine	0-0.15
p-OH benzoic acid esters	0-10
Lactic acid and its salts	no limit
Propionic acid and its salts	no limit
Natamycin/pimaricin	0-0.3
$NaNO_3$ and KNO_3	0-5
$NaNO_2$ and KNO_2	0-0.2
Sorbic acid including its Na/K/Ca salts	0-2.5
SO_2, Na_2SO_3, $NaHSO_3$, Na/K metabisulphite	0-0.7

The use of a particular additive in a food is allowed if the total daily intake through dietary consumption is less than the ADI which in actual practice is hardly 15 ± 5% of the ADI.

Economics of the additive Despite satisfying all the criteria mentioned in Table 2.1 an antimicrobial preservative may still not be widely used by the industry unless it is cheap enough, reduces spoilage and minimises food borne illness. For example, even though sorbic acid is far superior to benzoic acid in antimicrobial action as well as sensory quality, its use is less prevalent mainly due to its higher cost.

The subject of suitability and utility of a preservative is wide and has been dealt in detail later under respective heads.

Water Activity

The regulation of water as an ingredient of food was the earliest form of food preservation. The availability of water for microbiological growth and biochemical reaction can be controlled by dehydration, freezing or the addition of solutes such as salt and sugar. Water activity (Wa) is a measure of the availability of liquid water and is defined as the ratio of the equilibrium vapour pressure of the sample (P) to the equilibrium vapour pressure of pure water (Po) at the same temperature Wa=P/Po and can have values varying from 0 to unity. Equilibrium Relative Humidity (ERH) refers to the atmosphere surrounding the food and is equal to 100 x Wa where Wa refers to the activity of water in solid and liquid foods.

Determination

By measurement of freezing point depression or ERH can be carried out either statically by ERH measurement of the food in an enclosed space once moisture equilibrium has been attained or dynamically by passing a gas stream of known RH over the food and measuring the loss/gain of moisture by the sample.

Simple laboratory procedures for the determination of ERH include the use of salt impregnated filter paper and the non-equilibrium sorption rate method in which samples of food are placed in atmospheres of known ERH like saturated salt solutions. The food, depending upon its Wa will either absorb or lose moisture. By graphically plotting weight change after a fixed period of time against RH, the interpolated ERH at which no weight change occurs figures for Wa can be obtained. This method is best suited for relatively dry and solid foods with Wa < 0.90.

Several instrumental methods are available which find widespread use in the food industry. The simplest is the hair hygrometer which depends upon the change in length of a hair with change in RH. It can be used to measure the Wa of meat products at levels higher than 0.85. Psychrometers, vapour pressure manometers and dew point apparatus are also employed. Electronic hygrometers known for their convenience, accuracy and precision are widely used in food manufacture. These instruments incorporate sensors to detect the equilibrium humidity above the sample contained in a small chamber, maintained at a selected temperature. Sensors usually depend on substances like lithium chloride which show variations in conductivity with humidity. The electric signal created by the sensor is electronically amplified and Wa or ERH is presented as a recorded sorption isotherm or digital display. With different sensors Wa can be measured over the full range on all types of food using only a few grams of the sample.

The Pattern of Microbial Death

A microbial population is not killed instantly when exposed to a lethal agent. Population death like population growth, is generally exponential or logarithmic, i.e. the population is reduced by the same fraction at constant intervals as tabulated in Table 2.2.

Table 2.2

A theoretical microbial killing experiment

Minutes	*Microbial number at start of minute*	*Microorganism killed in 1 minute (90% of total)**	*Microorganism at end of 1 minute*	*Log_{10} of survivors*
1	10^6	9×10^5	10^5	5
2	10^5	9×10^4	10^4	4
3	10^4	9×10^3	10^3	3
4	10^3	9×10^2	10^2	2
5	10^2	9×10^1	10^1	1
6	10^1	9	1	0
7	1	0.9	0.1	-1

* Assuming that the sample contains 10^6 vegetative microorganisms/ml and that 90% of the organisms are killed during each minute of exposure.

When the population has been greatly reduced, the rate of killing may slow due to the survival of a more resistant strain of the microorganism.

In order to study the effectiveness of a lethal agent, it is necessary to ascertain whether the microorganisms are dead or not which is a very difficult task. It is impossible to take a bacterium pulse. A bacterium is defined as dead if it does not grow when inoculated into culture medium that would normally support its growth.

Conditions Influencing the Effectiveness of Antimicrobial Agent Activity

Destruction of microorganisms and inhibition of microbial growth is not an easy task because the efficiency of an antimicrobial agent, i.e. an agent that kills microorganisms or inhibits their growth is affected by at least following six factors:

(i) Population size: As an equal fraction of a microbial population gets killed during each interval, a large population requires a longer time to the than a smaller one irrespective of process applied or chemical antimicrobial agent used as shown in Table 2.2.

(ii) Population composition: The effectiveness of an agent varies greatly with the nature of the organism being treated because microorganisms differ markedly in susceptibility. Bacterial endospores are much more resistant to most antimicrobial agents than are vegetative forms, and younger cells are usually more readily destroyed than mature organisms. Some species are able to withstand adverse conditions better than others. *Mycobacterium tuberculosis*, which causes tuberculosis is much more resistant to antimicrobial agents than most other bacteria.

(iii) Concentration or intensity of an antimicrobial agent: Often, but not always, the more concentrated a chemical agent or intense a physical agent (thermal, irradiation, etc.) is the more rapid is the destruction of microorganisms. Dependence of effectiveness on concentration or intensity is generally not linear. Over a short range, a small increase in concentration leads to an exponential rise in effectiveness; beyond a certain point, further increase may not raise the killing rate at all. Sometimes, an agent is more effective at lower concentrations. For example, 70% ethanol is more effective than 95% ethanol.

(iv) Duration of exposure: Longer a population is exposed to an anti microbial agent, more are the organisms that get killed. To achieve sterilization, an exposure duration sufficient to reduce the probability of survival to 10^{-6} or less should be used.

(v) Temperature: An increase in the temperature at which a chemical acts often enhances its activity. Frequently, a lower concentration of disinfectant or sterilizing agent can be used at a higher temperature.

(vi) Local environment: The population to be controlled is not isolated, but surrounded by environmental factors that may either offer protection or aid in its destruction. For example, because heat kills more readily at an acid pH, acid foods and beverages such as fruits and tomatoes are much easier to pasteurise than more neutral foods like milk.

A second or important environmental factor is organic matter that can protect microorganisms against heat and chemical disinfectants. It may be necessary to clean an object before it is disinfected or sterilized. For example syringes and other equipment used should be cleaned before sterilization because the presence of too much organic matter could protect pathogens and increase the risk of infection. The same care must be taken when pathogens are destroyed during the preparation of drinking water.

SULPHURDIOXIDE (SO_2), SULPHITES AND BISULPHITES

In aqueous solution SO_2 exists as sulphurous acid (H_2SO_3), sulphite $(SO_3)^-$, and bisulphite ions $(HSO_3)^-$. The microbial growth inhibiting effects of SO_2 are most intense when it is in the unionised form as H_2SO_3. Also SO_2 is more inhibiting to bacteria than yeasts and molds. Major use of SO_2 and its salts as antimicrobial agents is in beverage and fruit preservation. Its use in dehydrated fruits and vegetables, soups and other mixes is mainly to control non-enzymatic browning. In beverages and fruits, SO_2 and its salts are primarily used to control acetic acid producing malolactic bacteria, fermentation and spoilage yeasts and fruit molds. Among the acetic acid producing bacteria *Acetobacter peroxydans* are most important. It has been observed that 150 ± 50 mg/litre SO_2 is essential for inhibiting *Acetobactor* cells in grape juice. Also common bacteria in acidic fruits and beverages are lactic acid producing genera, *Lactobacillus, Leuconostoc* and *Padiococcus.*

In wines, levels about 120 mg/litre of SO_2 decreased incidence of malolactic fermentation therefore it is widely used to deplete 'wild' yeasts in grape juice. Among the various yeast species, *Torulopsis* and *Saccharomyces* are most tolerant and are becoming a problem in juices, soft drinks and wine industry.

Mold infection during transit and storage is one of the serious problems in the marketing of fresh grapes and other fruits. *Botrytis sp* is probably the most prevalent fungi, but others infecting the fruits including *Cladosporium, Alternaria, Stemphyllium, Penicillium, Aspergillus, Rhizopus* and *Uncinula.* Slow release SO_2 generators "Grape guards" which can be placed into storage containers seem to be quite beneficial for the transport of grapes and are used commercially.

SO_2 is widely used as a temporary preservative in fruit pulps and juices in raw or semi-finished products and removed again in the course of processing during the action of heat or vacuum so that the residual quantities in the end product are minimised. Besides, antimicrobiological function in almost all food products where SO_2 is used, it has other technological functions to perform such as protection against oxidative, enzymic and non-enzymic browning reactions and inhibition of chemically induced colour and vitamin losses. The applied concentration of SO_2 necessary for these purposes are higher than the concentrations needed for microbiological control. Depending upon the products, in general between 0.01 and 0.2% SO_2 is employed and higher concentrations can be used in selected cases.

SO_2 reacts with vitamin B_1 and makes it unavailable. Therefore use of SO_2 should not be recommended for foods which serve as major sources of this vitamin in the diet.

SO_2 decolourises foods which derive their colour from anthocyanins.

Anthocyanins are pigments present in the sap of the plant cells in the form of glycosides and are responsible for the red, blue and violet colours of many fruits and vegetables. Most of the anthocyanins are derived from 3,5,7-7, trihydroxy flavylium chloride and the sugar moiety is usually attached to the -OH group on C_3. Some anthocyanins have been found also to contain, besides colours, additional components like organic acids and metals like iron, aluminium and magnesium.

Substitution of -OH and -OCH_3 groups influences the colour of the anthocyanins as shown by Bannerman. Increase in the number of -OH groups tends to deepen the colour to a bluish shade whereas increase in the -OCH_3 groups increases redness. Anthocyanins in solution create an equilibrium between the coloured cation R^+ or oxonium salt and the colourless pseudobase ROH, which is pH dependent.

$$R^+ + H_2O \qquad\qquad ROH + H^-$$

As the pH is raised, more pseudo base is formed and the colour becomes weaker. However, in addition to pH there are other factors also which influence the colour of anthocyanins, including metal chelation and combination with other flavonoids and tannins.

Anthocyanins get bleached by SO_2. It is a reversible process that does not involve hydrolysis of the glycosidic linkage, reduction of the pigment or addition of bisulphate to a ketonic, chalcone derivative. The reactive species is the anthocyanin carbonium ion (R^+) which reacts with a bisulphite ion to form a colourless chroman-2 (or 4) sulphonic acid (R-SO_3H), similar in structure and properties to an anthocyanin carbinal base (R-OH).

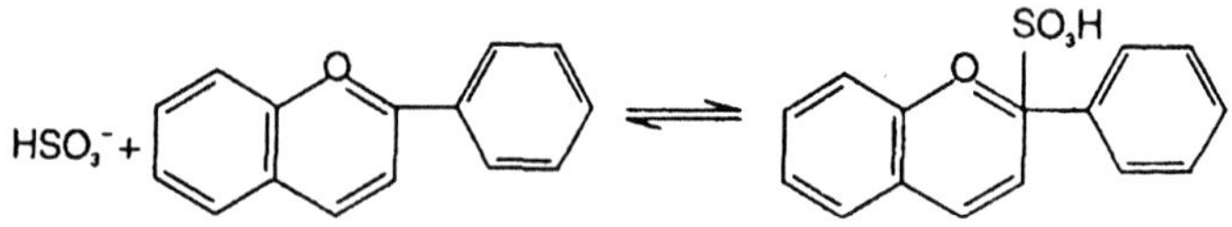

HSO_3^- + ⇌ SO_3H

Figure 2.1: Bisulphite reaction with anthocyanin carbonium ion

Recently, SO_2, sulphites and metabisulphites have been found to precipitate allergic reactions in about 10% asthmatic persons and their use in cut fruits has been banned in U.S.A.

Though under Rule 53(ii) (b) sulphurous acid including its salts have been permitted yet under Rule 55 all such additions shall be calculated as SO_2 and under column 3 the maximum permissible limits have been fixed only on the quantum of SO_2 estimated. This limit in various foods permitted for its use can be seen in Table 2.3.

Table 2.3

S. No.	Permitted article of food	Permitted quantity of SO_2 (ppm) (Max)
1.	Sausages and sausage meat containing raw meat, cereals and condiments	450
2.	Fruit, fruit pulp or juice (not dried) for conversion into jams or crystallised glace or cured fruit or other products	
a.	Cherries	2000
b.	Straw berries and raspberries	2000
c.	Other fruits	1000
3.	Fruit juice concentrate	1000
4.	Dried fruits	
	a. Apricots, peaches, apples, pears and other fruits	2000
	b. Raisins and sultanas	750
5.	Other non-alcoholic wines, squashes, crushes, fruit syrups, cordials, fruit juices and barley water to be used after dilution	
6.	Jams, marmalade, preserves, canned cherry and fruit jelly	40
7.	Crystallised glace or cured fruit (including candied peel)	150
8.	Fruit and fruit pulp not otherwise specified in the schedule	350
9.	Plantation white sugar, cube sugar, dextrose, gur, jaggery or misri	70
10.	Khandsari (sulphur) and Bura	150
11.	Refined sugar	40
12.	Corn flour and such like starches	100
13.	Corn syrup	450
14.	Canned *rossogullas (The can shall be internally lacqured with SO_2 resistant lacquer)*	100
15.	Gelatin	1000
16.	Beer	70

S. No.	Permitted article of food	Permitted quantity of SO_2 (ppm) (Max)
17.	Cider	200
18.	Alcoholic wines	450
19.	Ready to serve beverages	70
20.	Pickles and chutreys made form fruits or vegetables	100
21.	Dehydrated vegetables	2,000
22.	Syrups and sharbats	600
23.	Dried ginger	2,000
24.	Hard boiled sugar confectionery	350
25.	Dry mixes of Rasgollas	100
26.	Soups	
	(a) other than canned	150
	(b) Deried soups	1,500
	(c) Dehydrated soups mix, when packed in containers other than cans	1,500
27.	Fruits and vegetable, flakes, powder, figs	600
Appendix C Additions		
28.	(a) frozen shrimps (raws edible)	100
	(b) Shrimps (cooked product)	
29.	Synthetic syrups for dispensers	350
30.	Tomato Puree & paste	750
31.	Carbonated fruit beverages/drinks	70
32.	Dehydrated fruits	700
33.	Carbonated water, soft drinks, conc/liquid/ powder	70
34.	Fruit-based beverage Mix/powdered fruits-based beverages	120
35.	Candid crystallized & glazed fruit	150
36.	Murabba/Preserves	40
37.	Squashes/crushes/fruits-syrups, sharbats, cordial and barley water	350
38.	Ginger cock tail (ginger beer & ginger ale)	350
39.	(a) Fruit/Veg. Juice, pulp, puree with preservative for industrial use only	1000
	(b) cherry, strawberry, raspberry	2000

S. No.	*Permitted article of food*	*Permitted quantity of SO_2 (ppm) (Max)*
40.	Concentrated fruit/veg. Juice, pulp, puree with preservation for industrial use only	1,500
41.	Chutney fruits and/or vegetable/mango chutney	100
42.	Pickles	100
43.	Green chillies ginger/garlic/onion/whole chilli paste	100
44.	Carry over from fruit products	
	(a) Jams/jellies/fruit-cheese	40
	(b) fruit marmalades	40
	(c) fruit bar/toffee	100
	(d) thermally processed fruit beverage/drink/ ready to serve etc.	70
	(e) soup/fruit/vegetable/instant fruit/veg. chutney mixed dry/culinary/seasoning mixed powders	1,500
45.	Raisins	1500
46.	Grated desiccated coconut	50
47.	Dry fruit nuts	2,000
48.	Refined sugar	20
49.	Sugar icing/powdered sugar	20
50.	Dextrose	70
51.	Glucose syrup	40
52.	Glucose syrup (dried & meant for manufacture of confectionery to be sold	40
53.	Misri, gur/jaggery, plantation white sugar, cube sugar, golden syrup	70
54.	Khandsari sugar (sulphur sugar), bura sugar	150
55.	Cocoa powder	2000
56.	Chocolate-white milk, plain, composite, filled	150
57.	Sugar based/sugar free confectionery	2,000
58.	Chewing gum/bubble gum	2,000
59.	Lozenges	350

Estimation of SO_2

The presence of SO_2 is ascertained before commencing its estimation. A piece of filter paper dipped in lead acetate is dried. A few pieces of sulphur free zinc are added to a conical flask containing 10 to 15 gm of sample in solution and 20 to 25 ml of dil. HCl (1 + 3). The lead acetate paper is folded in a cone shape and placed into the mouth of the conical flask. The H_2S generated in the presence of sulphites turns the exposed portion of the lead acetate paper black.

Traces of metallic sulphides occasionally present in vegetables give the same reaction as sulphites under conditions of above test. Presence of SO_2 in such cases is to be confirmed by quantitative Monier Williams method described later on.

The available methods estimate the free or total SO_2. Free SO_2 is estimated by direct titration with iodine whereas total SO_2 is estimated by liberating the combined SO_2 by any of the following two methods,

(i) Treatment with excess alkali at room temperature, subsequent acidification to block recombination and iodo-metric titration.

(ii) Distillation from acid solution and titration.

Ripper modified the earlier method as described below:

Free SO_2

50 ml of sample (juice, squash, etc. diluted with water, if needed) is acidified with 5 ml of dil. H_2SO_4 (1 + 3). 500 mg Na_2CO_3 is added to expel the air if present, the contents titrated rapidly with 0.02 N standard iodine solution using starch as indicator. The blue coloured end point should persist for a few minutes. Let this titre value be *a*.

A similar aliquot of the substance is acidified with similar 5 ml of dil. H_2SO_4 (1 + 3), 10 ml of 38 ± 2% formaldehyde is added and kept aside for 10 minutes. The contents are rapidly titrated against the same 0.02 N iodine solution to faint but permanent blue colour. Let this titre be *b*.

Volume of iodine used by free SO_2 present in the sample

$$= a - b \ ml.$$

Total SO_2

A. To two similar aliquots of sample and 5 ml of 5 N NaOH is added to each. Contents stirred gently to avoid air introduction into the solution and kept aside for 20 minutes. To one of the sample 7 ml of 5 N HCl is added with stirring to avoid local concentration. 1 ml of 1% starch solution as indicator is added and the contents are titrated immediately with 0.02 N iodine to a definite dark blue colour employing a mechanical stirrer to check recombination. This titre c denotes the total iodine reducing value of the sample.

Reducing substances besides sulphite need to be determined, therefore the second sample is acidified, 10 ml of (38 ± 2%) formaldehyde added to bind the sulphite and kept aside for 10 minutes. Like before, using a mechanical stirrer for mixing, starch indicator is added and contents titrated rapidly against 0.02 N iodine solution till a dark blue colour persists for more than 15 seconds. Let this be *d*.

Volume of iodine used by the total SO_2 present in the sample $= c - d$ *ml.*

Calculation

1 ml of 0.02 N iodine = 0.64 mg of SO_2

SO_2 (ppm) = titre × 0.64 × 1000/wt. of sample

Combined SO_2 (ppm) = total SO_2 - free SO_2.

Total SO_2 is ascertained by distillation methods also. In one method the sample is acidified and the evolved SO_2 is swept with an inert gas like CO_2 or N_2 into cold H_2O_2 which oxidises H_2SO_3 to H_2SO_4. The later is titrated against a standard NaOH solution.

The second method is based on the distillation of SO_2 from the sample, its absorption in 0.1 N iodine solution and estimation by titration against 0.1 N $Na_2S_2O_5$ solution using 1% starch solution as indicator. *Not applicable for materials already containing sulphides.*

B. H_2O_2 method

Reagents

(i) 0.1% Bromophenol blue solution in water.

(ii) 0.1% phenolphthalein solution in 50% ethanol. 100 mg is dissolved in 50 ml of ethanol and then diluted to 100 ml with water.

(iii) Na_2CO_3—saturated solution prepared from (C.P.) quality and alkaline to phenolphthalein indicator.

(iv) 3% H_2O_2 solution is prepared by diluting 100 ml of 30% H_2O_2 to about 700 ml in a 1000 ml graduated cylinder. These contents are thoroughly mixed by pouring back and forth from a beaker to the cylinder. A 100 ml. portion from this solution is taken out separately and titrated in a 250 ml. beaker with 0.1N NaOH to pH 4.1 employing a pH meter. From this exercise the volume of NaOH required to neutralize the main solution is calculated. This test portion is discarded and the remaining main solution is neutralized to pH 4.1 as before.

H_2O_2 is standardized by pipetting 10 ml of the solution into a 100 ml volumetric flask and the volume made up. 5 ml of this diluted solution is pipetted in a 500 ml flask, about 300 ml H_2O + 10 ml of 6N H_2SO_4 are added and the contents titrated against 0.1N $KMnO_4$ solution to a permanent pink colour.

$$1 \text{ ml of } 0.1 \text{ N } KMnO_4 = 0.0017 \text{ g of } H_2O_2.$$

Apparatus See Fig. 2.2

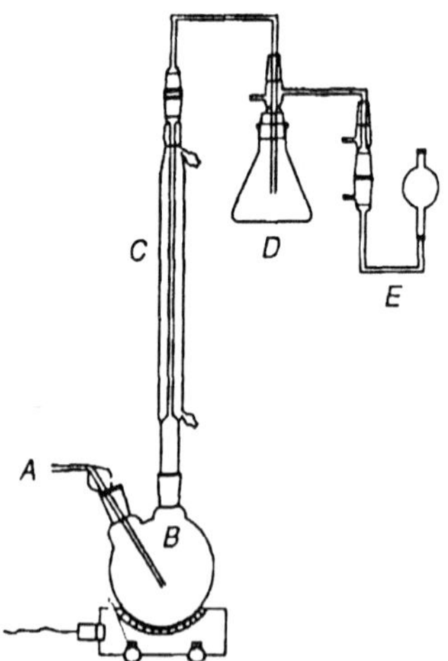

Apparatus
All glass distillation apparatus for determination of sulphur dioxide.
A–gas inlet tube, ***B***–500 ml round-bottomed flask, ***C***–condenser (400 mm jacket length), ***D***–250 ml conical flask, ***E***–trap.
The parts should have either 24/40 standard taper or 18/9 spherical joints.

Figure 2.2

Procedure

20 ml and 5 ml of 3% H_2O_2 are added to flask D and trap E respectively, cooling the condenser C by circulating cold water. 50 gm of homogenised sample is transferred through gas inlet into flask B using 300 ml of water, replacing the gas inlet A immediately and the entire apparatus is assembled, well greased and tightened. 20 ml of conc. HCl is slowly introduced through gas inlet and the same is again positioned and tightened. CO_2 washed through Na_2CO_3 solution or nitrogen gas (99.9% pure) @ 18 $\pm$ 2 bubbles/minute is passed through the gas inlet and flask B is heated to boil the contents first vigorously and then slowly. Only dried fruits and vegetables require boiling for an hour, the rest for 30 minutes. H_2O_2 solution from the trap is washed into flask D, the trap also rinsed with water. The contents are titrated against 0.05 N NaOH using 3 drops of bromophenol blue indicator. The end point is pale sky-blue. Type of burette and strength of NaOH are selected on the basis of expected SO_2 content in the sample. A blank on 20 ml H_2O_2 is done, correcting the net titre value accordingly.

Calculation

1 ml of 0.1 N NaOH = 3.2 mg of SO_2

$$SO_2 \text{ (ppm)} = \frac{\text{Titre} \times \text{normality of NaoH} \times 32 \times 1000}{\text{Wt. of sample}}$$

C. Accurately weighed 25 g of the food sample is pulverised and slurried in water. 20ml of conc HCl + 60+10 ml (as the flask permits) of boiled water is added through quick fit stop cock filter funnel. The liberated SO_2 is collected in a beaker already containing water. + a few drops of starch solution +0.1 ml of 0.02 N iodine solution in KI solution. Simultaneous titration is carried on against 0.02 N iodine solution till colour reappears.

1 ml of 0.02N iodine solution = 640 ppm of SO_2.

D. SO_2 bleaches p-rosaniline and by addition of formaldehyde, SO_2 forms an additional product releasing free rosaniline which is pink in colour. The amount of free rosaniline released is equivalent to the SO_2 present.

Reagents

(i) ***Formaldehyde solution (0.015%):–*** It is prepared by diluting 40% formaldehyde in two steps,

(a) from 10 to 1000 and (b) from 75 to 2000. i.e.

$$\frac{10 \times 75 \times 40}{100 \times 1000 \times 2000} = \frac{3}{2 \times 100} = \frac{3}{200} = 0.015\%$$

(ii) ***Acid bleached p-rosaniline hydrochloride:–*** 100 mg of p-rosaniline - HCl and 200 ml H_2O are taken in 1 litre volumetric flask. 150 ml HCl (1:1) is also added to the flask and the volume made up. For osmotic homogenisation, the flask is kept aside for 12 hours before use.

(iii) ***Sodium tetrachloromercurate:–*** 23.4 g NaCl + 54.3 g $HgCl_2$ are taken in a 2 litre volumetric flask, dissolved in 1000 ml of water and then diluted with water to volume.

(iv) ***SO_2 standard solution:–*** Approximately 170 mg $NaHSO_3$ is dissolved in water then diluted to 1 litre. It is standardised with 0.01 N iodine solution before use (approximately to yield 100 mg SO_2/ml).

Procedure

A standard curve is prepared by adding 5 ml (iii) to a series of 100 ml volumetric flasks followed by 0, 1, 2, 3 ml, etc. of (iv) in each flask. Contents of each flask is diluted with H_2O and homogenised. 5 ml portion from each flask is transferred to corresponding 200 mm test tubes containing, 5 ml (ii) then 10 ml of (i) is added, mixed and held for 30 minutes at 22°C. Absorbances are read against zero standard (blank) leading to plotting of the standard curve.

Estimation

10 ± 0.02 g ground dried fruit is blended with 290 ml of water for 2 minutes. 10 g aliquot is withdrawn from the bottom of the blender with a 10 ml calibrated free running pipette and transferred to a 100 ml volumetric flask containing 4 ml of 0.5N NaOH. Only 2 ml quantity is required for apples whereas only 1 ml shall suffice for golden raisins. Contents are swirled and mixed for about 25 ± 5 seconds. 4 ml 0.5 N H_2SO_4 (2 ml for apples and 1 ml for golden

raisins) + 20 ml (iii) are added and then diluted. Blank is also processed similarly but without 10 ml fruit extract.

2 ml sample solution is transferred to 200 mm test tube and procedure for the preparation of standard curve is followed. From the absorbance reading (SO_2 ppm) quantity is read correspondingly from the above standard curve.

If same colourimetric tube cell is used for successive samples it should be cleaned with 1:1 HCl and H_2O in between repeat use.

BENZOIC ACID

Benzoic acid occurs in nature in free and combined forms. Gum benzoin may contain upto 20%. Most berries contain appreciable amounts (0.05%). Its density is 1.321 (also reported as 1.266), mt. pt. 122.4°C, 2nd sublimation point is between 100°C, steam volatile. K at 25°C 6.40 x 10^{-5}, pH of saturated solution at 25°C 2.8, solubility in H_2O (at 0°C) 1.7 g/litre which increases with rise of temperature, at 10°C = 2.1, at 20°C = 2.9, at 25°C = 3.4, at 30°C = 4.2, at 40°C = 6.0, at 50°C = 9.5, at 60°C = 12.0, at 70°C = 17.7, at 80°C = 27.5, at 90°C = 45.5 and at 95°C = 68.0.1 gm dissolves in 2.3 ml cold alcohol, 1.5 ml boiling alcohol, also soluble in volatile and fixed oils. Solubility in water is increased by alkaline substances such as borax, trisodium phosphate.

Benzoic Acid and its Salts

In most countries benzoic acid and its Na/K salts are permitted for food preservation. The maximum permissible quantities vary from 0.15 to 0.25%. In USA, benzoic acid and Na benzoate are deemed GRAS (generally recognised as safe) upto a maximum concentration of 0.1%. In UK there is wide range. The permitted quantity of Benzoic Acid in the various food items in India is given in Table 2.4.

Table 2.4

S. No.	Permitted Article(s) of food	Permitted quantity of Benzoic Acid ppm
1.	Non-alcoholic wines, squashes, crushes, fruits syrups, cordials, fruit juices and barley water to be used after dilution	600
2.	Jam, hamalade, preserve, canned cherry and fruit jelly	200
3.	Ready to serve beverages	120
4.	Brewed ginger beer	120
5.	Coffee extract	450
6.	Pickles & chutneys made from fruits or vegetables	250
7.	Tomato and other sauces	750
8.	Danish tinned caviar	50
9.	Tomato puree & paste	750
10.	Syrups & Sharbats	600
11.	Fat spread, its Na/K salts, calculated as acid	1,000
12.	Sweet (carbohydrates based and milk product based) like – *halwa*, *Mysore pak*, Boondi Ladoo, Jalebi, Khoya burfees, peda, gulab Jamun, Rasgolla and similar milk products based sweets sold by any name	300
13.	Flavour emulsion/paste for carbonated and non-carbonated water only	GMP
14.	Ready-to-serve beverages	120
15.	Brewed ginger beer	120
16.	Tamarind pulp/puree & concentrate (acid, its Na/K salts or both calculated as acid)	750
17.	Synthetic syrups for dispensers (acid as above at S. No. 16)	500
18.	Tomato puree & paste (acid its salts as at S. No. 16)	120
19.	Carbonated fruit beverages/drinks	120

S. No.	Permitted Article(s) of food	Permitted quantity of Benzoic Acid ppm
20.	Carbonated water, soft drink conc. (liquid/powder)	120
21.	Murabba/Preserve	200
22.	Squashes, crushes, Fruit syrups, *sharbats*, cordials, and barley water	600
23.	Ginger cocktail (ginger beer & ginger ale)	600
24.	Fruit/Veg. Juice, pulp, puree, with preservatives for industrial use only (concentrated K/Ca/Salts & acid)	100
25.	Same as above but in concentrated form	600
26.	Chutney/Fruits and/or vegetable/ mango chutney	250
27.	Pickles	250
28.	Green chilli/ginger/garlic/onion/ whole chilli paste	250
29.	Jam/jellies/fruit cheese	200
30.	Fruit marmalades/bar/toffee	200
31.	Thermally processed fruit beverages/drinks/ready to serve beverages	120
32.	Culinary paste/other sauces	750
33.	Soya bean sauce	750
34.	Tomato ketchup	750
35.	Table olives	1,000
36.	Cocoa powder	1,500
37.	Chocolate-white milk, plain, composite filled	1,500
38.	Sugar based/free confectionery	1,500
39.	Chewing gum/bubble gum	1,500

Use of benzoic acid is mainly directed against control of yeasts and molds. Most yeast and fungi are inhibited by 0.05 to 0.1% of the undissociated acid. Bacteria are only partially inhibited, therefore, benzoic acid alone cannot be effective for preservation of food capable of supporting bacterial growth. Benzoic acid has been found to act synergistically with both sorbic acid and SO_2 and combinations of benzoic acid and sorbic acid have been reported to inhibit many bacterial strains better than either of these alone. Effectiveness of benzoic acid increases with lowering of pH and it is most effective below pH 4.5.

Benzoic acid is widely used in the preservation of fruit products and pickled vegetables in comparison to sorbic acid due to its lower price. Fruit pulps can be preserved by addition of 0.1 to 0.13% sodium benzoate against molds and fermentation so also fruit juices and ready to serve fruit beverages as evident from the above table. Combined with pasteurisation to inactivate enzymes and with small quantity of SO_2 as antioxidant it is applied in 0.05 to 0.2% concentrations depending upon the type of juice and length of time to which preservation is desired.

Sodium benzoate alone as well as with sorbate is used in the preservation of margarine, peanut butter, mayonnaise and fish marinades.

Rule 55 of the PFA Rules, 1955, restricts the use of class II preservatives in quantities mentioned in Table 2.4. There are natural products like the scent glands of the beaver, the bark of the wild black cherry tree, cranberries, prunes, ripe cloves and oil of aniseed which contain smaller amounts of free benzoic acid.[1]

These natural products like cranberries and prunes, i.e. dried plums are converted into jams and preserved by using ripe cloves and aniseed as ingredients for flavouring purposes, same is the case for *chutneys,* sauces, purees and pastes. However it has been observed that the content of benzoic acid is always slightly higher than the permitted quantity, as the rule restricts the use, i.e. extraneous addition and not the total content of benzoic acid.

Detection of Benzoic acid

$FeCl_3$ test The food sample is acidified with dil. HCl (1:3) and extracted with $(C_2H_5)_2O$. The solvent is evaporated on a hot water bath removing last traces of solvent under an air current. Residue is dissolved in a few ml of hot H_2O + a few drops of 0.5% $Fecl_3$, solution. Salmon colour precipitate of Fe benzoate indicates presence of benzoic acid.

Modified Mohler's test To the aqueous solution of the residue obtained as above, 1-2 drops of 10% NaOH solution are added and evaporated to dryness. To the residue 8 ± 2 drops of H_2SO_4 and a small crystal of KNO_3, contents heated for 10 minutes in a glycerol bath at 125 ± 5°C. After cooling, 1 ml water is added to make it distinctly ammonical. The solution is boiled to decompose any NH_4O_2 formed. After cooling, a drop of fresh and colourless $(NH_4)_2S$ solution is added for keeping the layers separate. Red brown ring indicates the presence of benzoic acid. On mixing, colour diffuses throughout the liquid and on heating finally changes to greenish-yellow. This change differentiates benzoic acid from salicylic acid or cinnamic acid, because they form coloured compounds which get destroyed on heating.

Estimation of Benzoic acid

(i) To a suitable volume of sample, 20 ml of buffer solution of 4.0 pH (containing 10% sodium citrate + 6.5% citric acid) is added. Benzoic acid is extracted with 3 x 25 ml of ether. Each extract is washed with 4 ml of water. The washed and combined extracts are filtered and the solvent is removed. Residual ether vapours are removed by air blowing. The residue is dissolved in 2 ml each of acetone and water and then titrated against 0.05 N NaOH, using phenolphthalein as indicator.

Calculation

1 ml of 0.05 N NaOH = 0.0061 g of benzoic acid.

(ii) In NaCl solution of the sample, the benzoic acid present is converted into water soluble sodium benzoate by the addition of NaOH. HCl is added to neutralize excess NaOH or to reform water insoluble benzoic acid which is extracted with $CHCl_3$. The $CHCl_3$ is removed by evaporation and the residue containing benzoic acid is dissolved in alcohol and then titrated with standard NaOH as above.

Calculation

1 ml of 0.05 NaOH = 0.0072 g of anhydrous sod. benzoate

Or

0.0061 g of benzoic acid as the case may be.

Preparation of sample

Solid or semisolid sample is homogenised. 100 g of it is taken in a 500 ml volumetric flask to which, NaCl powder is added for saturating the water in the sample. It is then made alkaline with 10% NaOH (checked by litmus paper). Contents are thoroughly shaken and kept aside for 2 hours with frequent shaking. Alternatively they are kept overnight for osmatic homogenisation and then filtered through Whatman No. 4 filter paper. Fat from fat rich samples passes through the filter paper into the filtrate. A few ml of NaOH is added for saponification and saponified fat is extracted with ether and separated from the defatted material.

Samples like cider which are having alcoholic content are made alkaline as above, contents evaporated on 250 ml steam bath till 100 ml is left. This residue is transferred into a 250 ml volumetric flask. 30 g of NaCl added, contents shaken for complete dissolution. After complete dissolution, volume made up with saturated salt solution is kept for at least 2 hours accompanied with shaking for homogenisation and is finally filtered.

In P.F. A. Act laboratories the following distinct procedures have been laid for sample preparation.

Beverages and liquid products The sample is mixed thoroughly. 100 g of it is transferred into a 250 ml volumetric flask, using NaCl brine. 10 ml of 10% NaOH solution is added and the volume made up with brine. The contents are thoroughly mixed and kept aside for 2 hours then it is filtered, this filtrate is used for benzoic acid determination.

Sauces and ketchups 100 g NaCl and 75 g of weighed sample are together quantitatively transferred to the 250 ml volumetric flask with brine. 10 ml of 10% NaOH solution is added, volume made up with brine and proceeded as above.

Ja.ns, jellies, preservatives and marmalades 75 g sample is weighed accurately, is mixed and transferred with 150 ml NaCl brine and then the previously mentioned procedure is followed.

Estimation

A. Titrimetric method 100 ml of the filtrate is pipetted into a 500 ml separating funnel. Dil. HCl (1 : 3) is added till it becomes neutral to litmus. 5 ml of excess HCl is added. 70,50,40 and 30 ml portions of $CHCl_3$ are successively used for extraction. Continuous rotatory shaking prevents emulsification and separates the $CHCl_3$ layer after being kept aside in a separating funnel. Entire clear $CHCl_3$ layer is carefully drawn off after each extraction which must be emulsion free. A H_2O wash of $CHCl_3$ layer is advantageous but not compulsory.

The combined $CHCl_3$ extracts from the separating funnel are transferred to a 250 ml dry conical flask rinsing the funnel with 10, 10 and 5 ml portions of $CHCl_3$. The contents are distilled very slowly at low temperature to about ¼ of the original volume. The residue is evaporated to dryness at ambient temperature in a current of dry air on a water bath till only a few drops are left. They are dried in a desiccator using H_2SO_4 as desiccant. Absence of acetic acid is ensured for products like ketchup containing acetic acid as an ingredient. The residue is dissolved in 50 ml of phenolphthalein neutral ethanol, 14 ± 1 ml of water is added along with an indicator and titrated against 0.05N NaOH.

Calculation

Anhydrous sod. benzoate (ppm) =

$$\frac{\text{Titre} \times \text{Normality of NaOH} \times 144 \times \text{made up volume} \times 10^6}{\text{Volume taken for estimation} \times \text{wt. of sample} \times 10^3}$$

Benzoic acid estimation calculation requires factor 122 in place of 144 in the above equation.

B. Spectrophotometric method Benzoic acid is extracted from separated sample using $(C_2H_5)O$ and the absorbance of the ether layer is measured at 272, 267.5 and 276.5 nm in the UV region. Amount of benzoic acid is determined from the corrected absorbance and the calibration graph obtained using standard benzoic acid solution.

Reagents

(i) $(C_2H_5)_2O$, i.e. diethyl ether
(ii) NaCl brine
(iii) NH_4OH(0.1%)
(iv) Dil. HCl (1 : 3)
(v) Standard benzoic acid.

Procedure

10 ml or 10 g of thoroughly mixed sample is transferred into a 200 ml separating funnel with NaCl brine, dil. HCl (iv) above mixed to make a litmus-acid solution.

Preparation of standard curve Absorbance of benzoic acid (AR) solution (50 mg/litre) is determined in a tightly stoppered cell between 265 and 280 nm at 1 nm intervals and plotted against wave length (WL). WL minimum at approx 267.5 nm is recorded as B. Minimum absorbance 'D' is at 276.5 nm (approx.) while highest absorbance 'C' is at approx. 272 nm.

20, 40, 60, 80, 100 and 120 mg/litre benzoic acid solutions are prepared in ether and their absorbances at point B, C and D are determined in the spectrophotometer. Average is obtained by subtracting absorbance concentration at B and D from absorbance at C. The difference plotted against concentration yields the standard curve.

Estimation The prepared solution is extracted with 70,50,40 and 30 ml portions of $(C_2H_5)_2O$. Emulsions formed are broken by standing/stirring/centrifuging and complete extraction is ensured by shaking well. The aqueous phase is drained and discarded. All ether extracts are combined, the end solution is washed with 40 and 30 ml portions of dil. HCl (1:1000) and HCl washings are discarded. If extraction does not need purification then the procedure detailed in the next para is adopted, otherwise the ethereal solution is extracted with 50, 40, 30 and 20 ml portion of 0.1% NH_4OH. The ethereal layers are drained and discarded. The alkaline extracts are neutralized with HCl + 1 ml excess. This acidified solution is again extracted with 70,50,40 and 30 ml portions of ether as mentioned earlier.

These ether extracts are combined and diluted to 200 ml with ether. Its absorbance is determined in tightly stoppered cells at wave lengths B, C and D after diluting with ether if necessary to obtain optimum concentration of 20-120 mg/litre. Absorbance at B and D are averaged and subtracted from absorbance at C. Concentration of benzoic acid is estimated from the standard curve making appropriate corrections for dilutions.

Calculation

Benzoic acid x 1.18 = sod. benzoate

C. ***Estimation of Benzoic acid in synthetic ready-to-serve beverages where saccharin is present*** Benzoic acid and saccharin are extracted together from the acidified beverages using $(C_2H_5)_2O$, the mixture is titrated with standard NaOH solution. Saccharin is estimated separately by colourimetric method and the titre equivalent to saccharin content in the sample is deducted from the total titre to calculate benzoic acid content of the sample.

Reagents

(i) Distilled $(C_2H_5)_2O$
(ii) Anhydrous Na_2SO_4
(iii) Dil. HCl (1 : 3)
(iv) 0.05 N NaOH
(v) Phenolphthalein indicator solution.

Procedure

25 g of homogenised beverage sample plus 10 ml of (iii) are mixed in a 250 ml separating funnel and extracted with 50,40,30 and 30 ml portions of $(C_2H_5)_2O$. The extracts are combined and then extracted with about 15 ml of water by gentle swirling to remove any traces of mineral acid. The-aqueous phase is discarded. The ethereal layer is passed through (ii) then desolventised on a water bath, and the last traces are blown out by air. The residue is dissolved in neutralized alcohol and titrated against (iv) using phenolphthalein as indicator. The titre (A) is the titre equivalent to the benzoic acid saccharin mixture.

Saccharin content (ppm) of the sample is determined colourimetrically as detailed under artificial sweeteners, say (S).

Calculation

$B = W \times S \times 10^{-6} \times 0.05/0.00916$,

where, B = Titre equivalent to saccharin content of the total sample, W = Wt. of the sample taken for estimation and Benzoic Acid Content = $A - B \times 0.05\ N \times 122 \times 10^3/W$.

Note: All notations have same meanings.

SORBIC ACID

Sorbic acid in the *trans, trans* isomer form is the commercial product, which needles out from water, mt. pt. is 134.5°C but it should be stored at temperatures below 40°C. Its solubility in water at 30°C is 0.25%, at 100°C is 3.8% in temperatures below 40°C. Its solubility in water at 30°C is 0.25%, at 100°C is 3.8% in propylene glycol, at 20°C is 5.5% in absolute ethanol or methanol 12.90%, in 20% ethanol 0.29%, in glacial acetic acid 11.5%, in acetone, 9.2%, in benzene 2.3%, in CCl_4 1.3%, in cyclohexane 0.28%, in dioxane 11.0%, in isoglycerol 0.31%, in iso propanol 8.4%, in isopropyl ether 2.7%, in methyl acetate 6.1%, in toluene 1.9%.

Sorbic Acid and its Salts

Sorbic acid and its salts are effective against moulds, yeasts and many bacteria. The growth of mycotoxin-forming moulds is permanently inhibited. There are no known reports of acquired resistance to sorbic acid. To ensure a reliable effect, sorbic acid and its salts must be used in certain minimum concentrations, which are dependent upon several factors like type of food, potential bacterial count, storage conditions, etc.

Bacteria are only partially inhibited; catalase positive are more sensitive than catalase negative. Microbial inhibitory action of sorbic acid is also pH dependent and decreases with rise in pH. However, even at higher pH, sorbic acid is more effective than propionates and benzoates and therefore it is widely used for bakery and confectionery products like cakes, fillings for chocolates and various types of cheese and cheese spreads. Action of sorbic acid against mycotoxin forming organisms is especially important. Sorbic acid has also been successfully used in the development of shelf stable, ready-to-eat *chapatties* in India and elsewhere.

In cheese industry, sorbic acid is applied and is directly added to fresh or processed cheese as sorbic acid or its K salt. Sorbate is also incorporated in the brine solution or acid for dusting of cheese surfaces. Fungi static wrappers containing Ca sorbate are also very effective for the prevention of fungal growth on cheese and bread. Sorbic acid treated wrappers are useful for the packing of smoked fish products.

Sorbates have been found useful for prevention of surface mold growth on hard sausages. Recently the use of K sorbate has been recommended along with small quantities of nitrite for the preparation of cured ham and bacon. Use of sorbate helps in the reduction of nitrite's use because of their involvement in the formation of nitrosamines which are reported to be carcinogenic. Inclusion of K sorbate in the curing mixture prevents the growth of *Clostridium* and the formation of *Botulinum* toxin.

Sorbates are also used in the preservation of fermented vegetable products and vegetables pickled in vinegar. Gherkins and olives in vinegar are protected by 0.1 to 0.2% K sorbate against film forming yeasts and mold attack. Presence of sorbate also inhibits lactic acid fermentation only slightly but suppresses the growth of film-forming yeasts and molds.

Sorbates have also been recommended for the preservation of prunes and fruit pulps. For this purpose, sorbates are combined with small quantities of SO_2 to prevent enzymic browning and oxidation. Sorbates at 0.02% have been suggested for the control of surface molds in jams and jellies and protection against spoilage by yeast in soft drinks. Use of sorbic acid has been found to be highly beneficial to sterilize wines against refermentation. A combination of 225 ± 25 mg K sorbate and 30 ± 10 mg free SO_2 provides wines with comprehensive protection.

Sorbic acid and sorbates are permitted in almost all the countries in 0.15 + 0.05 % concentration. In USA, sorbic acid and sorbates are deemed GRAS, i.e. (Generally Recognised as Safe). In UK and EEC countries, use of sorbic acid and K sorbate is allowed in a number of food products in concentrations upto 0.2%. Sorbates are one of the most widely investigated preservatives about their safety and based on these data WHO has prescribed highest ADI of 25 mg/kg body wt. for sorbates.

Under Rule 55 of P.F.A. Rules, 1955, sorbic acids and its salts have been permitted as described in Table 2.5.

Table 2.5

S. No.	*Articles of food Permitted*	*Types allowed*	*Max. permissible limit in ppm. of Sorbic acid*
1.	Cheese or processed cheese	Acid, its Na/K/Ca salts calculated as acid	3000
2.	Flour confectionery	Acid, its Na/K/Ca salts calculated as acid	1500
3.	Smoked fish in wrappers	Acid	Only wrappers may be impregnated with Sorbic acid
4.	Preserved chapatties	Acid	1,500
5.	Paneer or Chhana	Acid, its Na/K/Ca salts	2,000
6.	Fat spread	Acid, its Na/K/Ca salts Calculated as sorbic acid	1,000
7.	Jams, jellies, marmalades, preserves, crystallised, glazed or candied fruits including candied peels, fruit bars.	do	500
8.	Fruit juice concentrates with preservatives for conversion in juices, nectars for ready to serve beverages in bottles/pouches or selling through dispenser	do	100
9.	Fruit juices in tins, bottles or pouches	do	200
10.	Nectars, ready to serve beverages in bottles, pouches or selling through dispensers	do	50
11.	Prunes	K sorbate calculate as acid	1,000
12.	Bread & biscuit	like at (1) above	1,000
13.	Sweets (carbohydrates based and milk products based-like S.No. 12 of Tab 2.4.	Acid	1,000
14.	Flour confectionery	like at (1) above	1,500

S. No.	*Articles of food*	*Type allowed*	*Max. permissible limit in ppm.*
15.	Cake & pastries	-do-	1,500
16.	Salted fish	Na salt	200
17.	Candied crystalline and glazed fruit		500
18.	Murabba/Preserve		500
19.	Squashes, crushes, fruit syrups, sharbats, cordial and barley water		1000
20.	Ginger cocktail (ginger beer & ginger ale)		200
21.	Fruit/veg. juice pulp, puree with preservatives for industrial use only	Concentrated K/Ca/Salt & acid	100
22.	Chutney fruits and/or veg/ mango chutney	do	500
23.	Green chilli, ginger, garlic, onion, whole chilli paste	do	500
24.	Jams, jellies, fruit cheese, marmalades, bars, fruit toffees	do	500
25.	Thermally processed fruit beverages/fruit drinks/ ready to serve fruit beverages	do	300
26.	Tomato ketchup, culinary paste/other sauces and soya bean sauce	do	1000
27.	Nectars	do	50
28.	Table olives, dry fruits & nuts	do	500
29.	Cocoa powder	do	1,500
30.	Chocolate-white milk, plain, composite, filled	do	1,000
31.	Sugar based/sugar free confectionery	do	2,000
32.	Chewing gum/bubble gum	do	1,500

S. No.	*Articles of food*	*Type allowed*	*Max. permissible limit in ppm.*
33.	Cheese sliced/cut/shredded	do	3,000
34.	Processed cheese	do	3,000
35.	Processed cheese spread	do	3,000
36.	Cheddar cheese	do	1,000
37.	Extra hard grating cheese	do	3,000

The following list, arranged alphabetically, contains a selection of microorganisms, which are involved in decay processes in foods, against which sorbic acid is effective:

1. *Acetobacter aceti*
2. *Acetobacter xylinum*
3. *Achromobacter spp.*
4. *Acinetobacter*
5. *Aerobacter*
6. *Alcaligenes faccalis*
7. *Alternaria spp.*
8. *Asochyta spp.*
9. *Aspergillus elegans*
10. *Aspergillus flavus*
11. *Aspergillus fumigatus*
12. *Aspergillus niger*
13. *Aspergillus ochraceus*
14. *Aspergillus parasiticus*
15. *Aspergillus terreus*
16. *Aspergillus versicolour*
17. *Azotobacter agilis*
18. *Bacillus cereus*
19. *Bacillus coagulans*
20. *Bacillus subtilis*
21. *Botrytis cinerea*
22. *Brevibacterium*
23. *Campylobacter*
24. *Candida albicans*

25. *Cephalosporium spp.*
26. *Candida krusei*
27. *Candida tropicalis*
28. *Citrobacter*
29. *Clado sporium spp.*
30. *Clostridium perfringens*
31. *Clostridium sporogenes*
32. *Collectotrichum lagenarium*
33. *Corynebacterium*
34. *Cryptococcus spp.*
35. *Cryptococcus neoformans*
36. *Cryptococcus terreus*
37. *Debaryomyces spp.*
38. *Endomycopsis spp.*
39. *Enterobacter aerogenes*
40. *Epicoccum*
41. *Escherichia coli*
42. *Escherichia freundii*
43. *Flavobacterium*
44. *Fusarium episphaeria*
45. *Fusarium monoliforme*
46. *Fusarium oxysporum*
47. *Fusarium roseum*
48. *Fusarium rubrum*
49. *Fusarium solani*
50. *Fusarium tricinctum*
51. *Geotrichum candidum*
52. *Gluconobacter*
53. *Hansenula anomala*
54. *Hansenula saturnus*
55. *Hansenula subpelliculosa*
56. *Heterosporium terrestre*
57. *Klebsiella*
58. *Lactobacillus brevis*
59. *Leuconostoc mesenteroides*

60. *Micrococcus spp.*
61. *Mucor silvaticus*
62. *Mybacterium*
63. *Myrothecium roridum*
64. *Myrothecium verrucaria*
65. *Myrothecium spp.*
66. *Oospora spp.*
67. *Pediocuccus*
68. *Penicillium chermesinum*
69. *Penicillium chrysogenum*
70. *Penicillium citrinum*
71. *Penicillium digitatum*
72. *Penicillium frequentans*
73. *Penicillium funcicolosum*
74. *Penicillium italicum*
75. *Penicillium notatum*
76. *Penicillium patulum*
77. *Penicillium piscarium*
78. *Penicillium restrictum*
79. *Penicillium roquefortii*
80. *Penicillium regulosum*
81. *Penicillium sublateritium*
82. *Penicillium thomii*
83. *Penicillium urticae*
84. *Penicillium variabile*
85. *Penicillium spp.*
86. *Phoma spp.*
87. *Pichia polymorpha*
88. *Pichia silvestris*
89. *Pichia spp.*
90. *Propionibacterium Zeae*
91. *Propionibacterium freundenreichii*
92. *Protens vulgaris*
93. *Pseudomonas spp.*
94. *Pullularia pullulans*

95. *Rhizoctonia solani*
96. *Rhizopus nigricans*
97. *Rhodotorula flava*
98. *Rhodotorula glutinis*
99. *Rhodotorula rubra*
100. *Rhodotorula spp.*
101. *Rosellinia spp.*
102. *Sacchromyces carlsbergensis*
103. *Sacchromyces cerevisiae var ellipsoideus*
104. *Sacchromyces cerevisiae*
105. *Sacchromyces delbrueckii*
106. *Sacchromyces fragilis*
107. *Sacchromyces lactis*
108. *Salmonella heidelberg*
109. *Salmonella montevideo*
110. *Salmonella typhimurium*
111. *Salmonella enteritidis*
112. *Sarcina lutea*
113. *Schizosacchromyces octosporus*
114. *Serratia marcescens*
115. *Shigella*
116. *Staphylococcus aureus*
117. *Streptococcus*
118. *Streptomyces*
119. *Torulopsis Candida*
120. *Torulopsis caroliniana*
121. *Torulopsis minor*
122. *Trichoderma viride*
123. *Trichomonas*
124. *Vibrio parahaemolyticus*
125. *Yersinia*
126. *Zygo Sacchromyces bailü*
127. *Zygo Sacchromyces globiformis*
128. *Zygo Sacchromyces holomembranis*
129. *Zygo Sacchromyces rouxii*

Estimation of Sorbic acid

Sorbic acid is distilled in a two-neck distilling flask having round glass joints. A 125 ml separating funnel is positioned in one neck and in the other neck is fixed Friedrich's condenser. Interfering substances in the distillate are separated on a 1.5 cm x 25 cm chromatographic tube. Absorbance of thus isolated extract containing sorbic acid is measured at 254 nm.

Reagents

(i) Sorbic acid standard solutions 50 mg of sorbic acid is dissolved in isopropanol in a 500 ml volumetric flask and the volume made up. 5 ml of this solution is further diluted to 250 ml to get the working standard, therefore 1 ml of standard solution = 0.1 mg and 1 ml of working standard = 0.002 mg.

(ii) Bromocresol indicator 0.25% in methanol.

(iii) Solvents (a) 70% methanol.
(b)* $CHCl_3$: iso-octane :: 1 : 4,
saturated with (a) above.

100 ml $CHCl_3$ + 400 ml of iso-octane are mixed in a 1 litre separating funnel, 20 ml (a) added, contents are shaken vigorously, kept aside, methanol layer drained off. $CHCl_3$-iso-octane layer is filtered quickly through a fluted filter paper.

(iv) Silicic acid (Chromatographic grade)

Procedure

2.0 mg sorbic acid containing aliquot of liquid sample is weighed and transferred to the 2 neck distillation flask with water to a total volume of 50 ml.

In case of solid samples, 20 mg sorbic acid portion is blended with water and transferred to a 500 ml volumetric flask containing volume made up with water, homogenised and filtered. 50 ml of the filtrate is pipetted into the above distillation flask.

After the addition of 50 g of Mg SO_4, distillation is carried out and 40-45 ml of distillate is collected in a measuring cylinder. This distillate is transferred to a 250 ml separating funnel, rinsing the cylinder with hot H_2O. The washings too are transferred to the funnel. Contents are cooled, saturated with NaCl. Sorbic acid is extracted with 5 x 40 ml portions of $CHCl_3$, transferring each $CHCl_3$ layer to another separating funnel having 10 ml H_2O. Contents of the second separating funnel are passed through a wad of cotton. This filtrate is evaporated on a steam bath under an air current till the residue is about 25 ml. This residue is transferred to a 50 ml beaker and evaporated as before to about 10 ml. This portion of $CHCl_3$ is evaporated in an air current, keeping the temperature below 37.5°C. The residue is dissolved in 2 ml of (iii) (b).*

5 gm of (iv), 1 ml of (ii), 1 drop of IN NH_4OH and (iii) (a) is taken in a mortar in a quantity that (iv) can hold without becoming sticky when slurried with (iii) (b). Contents are thoroughly mixed, enough (iii) (b) is added, first to a paste and then a slurry. A thin wad of cotton is put at the constricted end of the tube attached to the column, slurry poured over it, excess solvent removed either by mild suction or air pressure (6+1 lbs). The column below gel surface is kept moist.

2 ml of sorbic acid extract is transferred to the ready column from the beaker. The beaker is washed with 3 x 2 ml of (iii) (b) and the washings are added to the column only after extract and each washing has passed into the column. When the last portion of the washing has passed into the column, (iii) (b) solvent is added and allowed to percolate through the column either under pressure or mild suction. The initial acid is discarded and sorbic acid band is collected in a 50 ml volumetric flask. After its complete collection the volume is made up with (iii) (b) solvent.

An aliquot containing 18 ± 2 mg of sorbic acid is pipetted into a 100 ml volumetric flask, the volume made up with iso-propanol and contents homogenised. Absorbance of the sample and the standard is measured at 254 nm spectrophotometrically.

Calculation

Mg of sorbic acid per 100 g. (liquid sample) =

$$\frac{0.2 \times \text{OD of sample} \times 100 \times 50 \times 100}{\text{Wt. of sample} \times \text{OD of standard} \times \text{volume taken* for final dilution}}$$

In the case of solid samples, weight taken for blending, volume made up, and aliquot taken for estimation should be included.

Detection of Benzoic Acid and Sorbic Acid
(TLC Method)

Under Rule 55 of PFA Rules, only fat spreads have been allowed to use sorbic or benzoic acids and their salts together. In such cases, these two preservatives are separated by steam distillation, extracted into ether from the acid solution and ethereal extract examined by TLC.

Reagents

(i) HCl 50% (v/v) (ii) $(C_2H_5)_2O$, peroxide free
(iii) C_2H_5 OH : NH_3:: 9 : 1 as TLC developing solvent
(iv) 1% standard solution of benzoic and sorbic acid in ethyl acetate
(v) $FeCl_3$: H_2O_2:: 1 : 1. $FeCl_3$ 2% freshly prepared and H_2O_2 0.5%
(vi) TBA spray agent made from 0.2% 2-thio barbituric acid in H_2O
(vii) $MgSO_4.7H_2O$
(viii) H_2SO_4(1M)
(ix) NaOH(1M)
(x) TLC plates coated with silica gel G.

Procedure

A weighed portion of sample, 100 g (vii), 100 ml (viii) is mixed and quickly steam distilled, collecting steam distillate in a flask already having about 15 ml of (viii). 3 × 2 5 ml of (ii) is used to extract about 100ml of distillate. The combined solvent layer is dried over anhydrous Na_2SO_4 and then reduced to 1ml by freeze drying system.

Not more then 20 ml is spotted on (x) above along with (iv). The chromatogram is developed upto 10 cm using, (iii). The plates are air dried then sprayed with (v). Benzoic acid shows as a mauve coloured spot (Rf. 0.5) and sorbic acid may be distinguished as a yellow coloured spot slightly below it. Further spraying with (vi) and heating at 100°C for 5 minutes, makes sorbic acid appear as a pink spot (Rf. 0.45).

Estimation of Sorbic acid

A. UV spectrophotometric method Sorbic acid is extracted from the sample by $(C_2H_5)_2O$:Pet. ether::1:1, absorbance of the extract is measured at 250 nm.

Reagents

(i) m-phosphoric acid; 5 gms is dissolved in 250 ml H_2O then diluted to 1 litre with ethanol.

(ii) Pet. ether:anhydrous $(C_2H_S)_2O$ :: 1 : 1

(iii) $KMnO_4$ solution (15%)

(iv) Sorbic acid:

(a) Standard stock solution (1 mg/ml) in (ii) above.

(b) Working standard solution; 5 ml of (a) is diluted to 100 ml with (ii) as before.

(v) Reference solution: 10 ml of (i) and (ii) are mixed by shaking, the supernatant layer is dried with anhydrous Na_2SO_4.

Procedure

The sample is homogenised. Cheese and related products are cut into small pieces with a food chopper or shredded over a sieve. About 10 g of sample and (i) are mixed to yield a total of 100 ml of liquid in the mixture. It is blended for 1 minute and filtered *immediately* through Whatman No. 3 filter paper 10 ml of this filtrate is transferred into a 250 ml separating funnel already containing 100 ml of (ii) and contents shaken for 1 minute. The aqueous layer is drained and discarded while the ether extract is dried our 5 g of anhydrous Na_2SO_4. Absorbance at 250 nm is read against (v).

1,2, 4 and 6 ml of (iv)(b) are taken in separately marked 100 ml volumetric flasks and the volume in each flask is made up with (ii) above, homogenised. Absorbance of each solution is determined at 250 nm and a standard curve plotted. From this curve the sorbic acid content of the sample is estimated.

Presence of sorbic acid is ascertained by adding 2 ml of (iii) to the remaining ether solution and mixed by shaking. The ethereal layer is filtered through No. 1 Whatman filter paper, the filtrate dried over anhydrous Na_2SO_4, absorbance readings of thus dried filtrate are taken between 220 and 300 nm. Absence of peak at 250 nm confirms the presence of sorbic acid.

B. Colourimetric method Acidified sample containing sorbic acid is steam distilled, the distillate oxidised with $K_2Cr_2O_7$ to melon aldehyde which is coupled with thio barbituric acid to give a red coloured solution which is absorbed at 532 nm.

Reagents

(i) H_2SO_4 (2N): 14.2 ml of H_2SO_4 is diluted to 50 ml with H_2O.

(ii) H_2SO_4 (0.3N): 15 ml of (i) is diluted to 100 ml.

(iii) $K_2Cr_2O_7$ solution: 147 mg of this reagent (AR) is dissolved in H_2O and diluted with H_2O to 100 ml.

(iv) TBA (Thiobarbituric Acid) solution (0.5%): 250 mg of TBA are dissolved in 5 ml of 0.05N NaOH in a 50 ml volumetric flask, swirling the flask under hot water. About 20 ml H_2O is added and the resulting solution is neutralized with 3ml of 1N HCl, the volume is then made up with H_2O. This is always freshly prepared.

(v) Sorbic acid standard solution: same as (iv) in method A above.

Procedure

Preparation of standard curve 5,10 and 15 ml of (v) are taken into separately marked 500 ml volumetric flasks making up the final volume with water. 2 ml from each above solutions + 2 ml H_2O (blank) is pipetted into separate 15 ml test tubes. 1ml each of (ii) and (iii) above is added to each test tube, each tube is heated on a boiling water bath exactly for 5 minutes then cooled in cold water, 2 ml of (iv) added. Tubes are again placed in boiling-water bath for 10 minutes, recooled. The absorbance of each tube is determined at 532 nm against a similarly prepared blank and from these a calibration graph is prepared.

Estimation

1.75 + 0.25 g sample is weighed into a distilling flask containing silica chips. 10 ml of (i) and 10 g of $MgSO_4 . 7H_2O$ is added and the contents are subjected to steam distillation maintaining 25 ± 5 ml volume in the distillation flask with a burner under it. Charring is avoided. 100-125 ml distillate is collected in a 250 ml volumetric flask in about 45 minutes. With 2 ml distillate the procedure is run as under standard curve and the quantum of sorbic acid is determined.

If the sample contains 0.05% sorbic acid, it is diluted to equivalent concentration.

NITRITES AND NITRATES

Nitrates were used for centuries in food preservation mainly for curing and pickling of meat. The discovery that nitrite was the active compound was made in 1890. It is now considered that nitrate is not an essential component in curing mixtures and therefore it is recommended sometimes that nitrate may be transformed into nitrite. Both nitrate and nitrite have antimicrobial action like in the production of Gouda cheese to prevent gas formation by butyric acid forming bacteria.

The action of nitrite in meat curing is considered to inhibit formations of toxins by *Clostridium botulinum* and *Staphylococcus aureus*. This is an important factor in establishing safety of cured meat products. The preservative action of nitrites is mainly due to the formation of nitrous acid and other oxides of nitrogen and their action increases with decreasing pH value. In meat industry, nitrites and nitrates are mainly used as colour fixatives. Nitrites react with myoglobin forming nitrosomyoglobin which has bright red colour and is not affected by atmospheric oxidation. Nitrites also delay off-flavour development during storage of processed meat products.

It has been observed that secondary amines in foods may react with the generated nitrous acid to form nitrosamines as explained in the following equation.

$$\begin{matrix} R_1 \\ R_2 \end{matrix} \Big> NH + HNO_2 \rightarrow \begin{matrix} R_1 \\ R_2 \end{matrix} \Big> N\text{-}N = O + H_2O$$

Nitrosanime

Nitrosamines have been reported to be carcinogenic in nature and these maybe both mutagenic and teratogenic. Since the quality of nitrosamines produced is of the order of ppm or even ppb, their precise estimation is not yet standardised, therefore, a clear picture has not emerged about their aforesaid implication. Besides, no suitable substitute for nitrites has been established and recommended for the production of ham, bacon and other cured meats. The ADI for nitrites is up to 5 mg/kg body wt/day, therefore a maximum concentration of 200 ppm has been allowed. It is calculated as $NaNO_2$ in cooked pickled meat including ham and bacon under Rule 55 of PFA Rules 1955 at S.No. 21 of the relevant schedule.

S. No	*Permitted Articles of food*		*Permitted Max limit quantity of NO_3^-/NO_2^- ppm*
1.	Pickled meat and beacon	Na and/or KNO_2 expressed as $NaNO_2$	200
2.	Corned beef	do	100
3.	Luncheon meat, cooked ham, chopped meat, canned mutton chicken, goat meat	do	200

The pigment of cooked meat, called hemochrome, is brown in colour. In the presence of reducing substances likely to be present in the interior of cooked meat, the iron present in the ferric (Fe^{+++}) state gets reduced to ferrous state (Fe^{++}), the resulting pigment is pink hemochrome. In the curing of meat heme reacts with the nitrite of the curing mixture. The nitrite heme complex called nitrosomyoglobin has a red colour but is unstable. On heating the more stable hemochrome-cured meats major pigment is formed and the globin portion of the meat gets denatured at 65°C.

The first reaction of the nitrite with myoglobin is the oxidation of Fe^{++} to Fe^{+++} and formation of met myoglobin. Simultaneously nitrate is formed.

$$4MbO_2 + 4NO_2^- + H_2O \rightarrow 4\ \text{Met MbOH} + 4NO_3^- + O_2$$

This explains the gradual lowering of nitrite content during the formation of cured pigment.

Nitrites are added to vat milk (0.01-0.02%) to avoid blowing in hard cheese.

Under Rule 55B: No $N0_2^-$ or $N0_3^-$ shall be added to any infant food.

Determination of Nitrate and Nitrite in Foods

The sample is clarified with alumina cream and the amount of nitrate present is determined by allowing it to diazotise arsenilic acid and coupling the diazonium salt with n-1-napthylethylene diamine. The colour so formed is extracted into n-butanol and its absorbance is measured at 545 nm. An aliquot of the sample is mixed with spongy cadmium (Cd) in order to reduce any nitrate present and the nitrite so produced is determined in the same way. The amount of nitrate present is then calculated by subtracting the nitrite from the total.

Reagents

(i) Water: This may be distilled or de-ionised but a blank must be carried out to check its suitability for the preparation of sponge Cd.

(ii) Alumina Cream: To a saturated solution of $KAlSO_4$, NH_3 is added slowly stirring until pH reaches 7.0.

(iii) n-napthylethylene diamine (0.1%) in H_2O.

(iv) Arsenilic acid monohydrate (0.1%) in 5M HCl

(v) Spongy Cd: Zinc rods are placed in 20% aqueous Cd SO_4 solution for about 4 hours, precipitated Cd is separated, washed twice with H_2O then macerated in it for about 3 hours. It is activated by shaking it with 2M HCl, washed at least 5 times with H_2O then finally kept under H_2O. For each batch fresh preparation is done.

(vi) Standard (NO_2^-) solution: 0.4783 gm $NaNO_2$ is dissolved in H_2O and 1 litre volume is made. 10 ml of this is further diluted to 100 ml to get 10 mg/litre of nitrite nitrogen.

Procedure

A macerated and homogenised 5 gm of the sample is weighed into a 50 ml beaker, 50 ml H_2O is added and it is heated to 80°C, stirring gently and maintained at 80°C for 10 minutes. 20 ml (ii) is added, contents are gently transferred to 100 ml volumetric flask, cooled and diluted to volume with water. After thorough mixing the entire volume is filtered through Whatman No. 4 filter paper rejecting the first 10 ml of filtrate.

Small amounts of nitrate are present in these filter papers hence the filter paper must be washed before hand with at least 100 ml of hot water.

(a) ***NO^- Determination:*** 10 ml of filtrate is pipetted into a 50 ml volumetric flask, 2 ml of (iv) is added and kept aside for 5 minutes, 2 ml of (iii) is added, mixed and kept aside for 10 minutes. Only a clear solution is diluted to 50 ml with H_2O and its absorbance at 538 nm is noted. Cloudy solutions are transferred to a 100 ml separating funnel, saturated with salt and extracted with n-butanol using 20, 15 and 5 ml portions. The butanol extract is passed through a small cotton pledget in a funnel and then into a dry 50 ml volumetric flask where it is diluted to volume with n-butanol. Its absorbance at 545 nm is read.

(b) ***NO_3^- Determination:*** 10 ml of filtrate is pipetted into a small stoppered conical flask, 5 ml of (v) and 1 g of wet Cd (vi) is added. The flask is stoppered and shaken for 5 minutes. This solution is filtered through a washed filter paper into a 50 ml volumetric flask, Cd and filter paper is also washed. NO_2 content is determined in the filtrate as given above starting at "2 ml of (iv) added".

(c) ***Preparation of Standard Curve:*** Into a series of 50 ml volumetric flasks dilute solution of $NaNO_2$ (vii) is pipetted. The flasks contain 2 to 15 μg of NO_2^- nitrogen and the colour is developed as at (a) above for nitrite. Absorbance is read and a standard curve is plotted. The experiment is repeated and the colour is extracted with n-butanol. The absorbances are read at 545 nm and the new standard curve is plotted.

From the graph, the nitrite content is calculated before and after reduction, and the NO_3^- content is calculated by subtraction.

ESTERS OF P-OH BENZOIC ACID, SYN. PARABENS

The esters of p-OH benzoic acid were produced mainly to replace benzoic acid and salicylic acid which were mainly effective in highly acidic products. CH_3-, C_3H_7-, C_4H_9-, C_7H_{15}- esters of p-OH benzoic acid despite having better antimicrobial activity than the parent acid have found less acceptance in food preservation due to their very low solubility and not so good organoleptic properties. Their antimicrobial action corresponds directly to chain length of the alcohol component, i.e. it increases with the chain length. Their action is fungistatic or similar to benzoic and sorbic acid. The phenolic group makes them more effective then benzoates against bacteria in general, gram-positive stains in particular. Minimum inhibitory concentrations against various bacteria and fungi at pH 6+1 varies from 12 to 400 ppm.

But their pronounced taste restrains their wide use. In India only CH_3- and C_3H_7 derivatives have been allowed and that too not more than 500 ppm in flour for baked goods. In USA and UK use of C_2H_5

ester has been allowed upto 1000 ppm. Its main use in these countries is for sausage preservation as well as meat coatings containing gelatin, fish marinades in combination with sorbic and benzoic acids. Alcoholic solutions of esters are occasionally used for preserving cheese and confectionery fillings to stop microbial growth. C_2H_5 ester is excellent for beer preservation hence has been allowed in USA.

These esters have very good efficiency in preventing microbial activity even in weakly acidic or neutral pH range. For this reason these esters have found major use in the preservation of cósmetics and pharmaceuticals.

C_4H_9-ester also known as n-butyl p-hydroxy benzoate, Butoben, Butyl Chemosept, Butyl Parasept, Tegosept B, is a crystalline powder having mt. pt. 68-69°C. It is very slightly soluble in water, (1:6500) and glycerine and is freely soluble in acetone, alcohol, ether, chloroform and propylene glycol.

$C_4 H_5$-ester is known as Nipagin A, Ethyl Parasept and Solbrol. It is crystal in form having mt. pt. 116°C, b.pt 297-298°C but decomposes and is freely soluble in alcohol and ether. Its solubility in water at 20°C is 0.070% (w/w) and at 25°C it is 0.075% (w/w).

$C_3 H_7$-ester is also known as propylparaben, Nipasol, Chemocide PK, Propyl Chemosept, Solbrol P, Propyl Parasept. It appears as white crystals having mt. pt. 96-97°C. Soluble in 2000 parts water and is freely soluble in alcohol, ether, slightly soluble in boiling water.

$C H_3$-ester known as methylparaben, Nipagin M, Tegosept M, Methyl Chemosept or Parasept appears as white needles having mt. pt 131°C, b.pt. 275 ± 5°C with decomposition. 1 g dissolves in 400 ml water, 40 ml warm water and about 70 ml warm glycerine; it is freely soluble in alcohol, acetone, ether. Its solubility in water at 20°C is 0.25% (w/w) and at 25°C is 0.30% (w/w).

All these have the following general configuration.

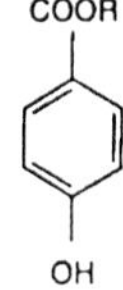

Where R may be CH_3-, $C_2 H_5$-, $C_3 H_7$-, $C_9 H_9$- etc.

Detection of Parabens and p-OHBenzoic Acid

A. TLC Method: The sample is acidified and extracted with $(C_2H_5)_2O$ and the extract is concentrated and subjected to TLC.

Reagents

(i) Silica gel G
(ii) Developing solvent : toluene : methanol : acetic acid :: 45:8:4
(iii) 2% solutions of acid and parabens.
(iv) Denige's Reagent: It is prepared by mixing 5 gm HgO and 40 ml of water, cooling in ice-salt mixture, adding freezing cold 20 ml of H_2SO_4 very slowly and stirring the contents.
(v) H_2SO_4 (10% v/v)
(vi) $NaNO_2$ 2%, freshly prepared.
(vii) Na_2SO_4 - anhydrous
(viii) Solvent ether.

Procedure

5 ml (v) is added to 10 gm of sample and ground with (vii) in a mortar until the sample is dry. Further grinding is done with small successive quantities of (viii) and ether is decanted. The ether extract is filtered and the filtrate is evaporated at a low temperature. The residue is dissolved in 1 ml of methanol.

20 μl is spotted and chromatogram is developed with (ii). The plate is viewed under uv (254 nm). Parabens show black spots and the area is marked with a pin and sprayed with (iv). Paraben gives a white spot, visible by different reflectivity from the background. After heating at 100°C for 5 minutes it is sprayed with (vi). Appearance of red spots indicate the presence of parabens.

B. Qualitative Test for 4-OH Benzoic Acid: The test is done on neutral ammonium salt. The food if not acidic is acidified and then 4-OH benzoic acid is extracted with ether and desolventised. The residue is dissolved in few drops of NH_4OH solution in a test tube.

Millon's Reagent is prepared by dissolving 3 ml mercury in 27 ml cold fuming HNO_3 and diluted with an equal volume of H_2O. A few drops of this reagent are added to the above solution. Presence 4-OH benzoic acid is revealed by rose-red colour.

Note: Many aromatic substances with a OH group attached to benzene ring, like salicylic acid, give tinted red colour, orange in the aforesaid case.

Estimation of Paraben(s): Parabens present in the sample are hydrolysed with alkali and the released p-OH benzoic acid is extracted with $(C_2H_5)_2O$ after acidification of the sample. After re-extraction with NaOH from ether, colour is developed with Denige's reagent and the absorption is read at 518 nm.

Reagents

(i) Denige's reagent: 5 g of HgO in 20 ml of conc. H_2SO_4 and diluted to 100 ml with H_2O.

(ii) $K_4Fe(CN)_6$ 15% in H_2O.

(iii) $ZnSO_4$ 30% in H_2O.

(iv) NaOH 5% in H_2O.

(v) $NaNO_2$ 2% freshly prepared in H_2O.

(vi) Dil. H_2SO_4; 100 ml conc. H_2SO_4 in 300 ml H_2O.

Procedure

To 2 g of the sample 60ml H_2O at 50°C is added. The pH of the contents is adjusted to 7.5 with (iv) and is heated at 50°C for 30 minutes with occasional stirrings. 2 ml of (ii) is added and mixed carefully. 2 ml of (iii) is mixed, the volume made up to 100 ml and then filtered. 50 ml of this filtrate and 1 ml (vi) are mixed and then extracted with 3 x 50 ml portions of $(C_2H_5)_2O$. The combined ethereal extracts are washed with H_2O (3x5 ml/30 seconds), a drop of phenolphthalein is added and then shaken with 3ml of 0.25 M NaOH solution. After washing with 3ml of H_2O, the alkaline extracts are combined, traces of ether, if any, are removed on a hot-water bath and the volume is made up to 10 ml. 5 ml of this solution and 5ml of (i) are mixed and heated in a boiling water bath for 5 minutes, cooled, then 5 drops of (v) are added and contents left aside for 45 minutes. A pink colour develops. The absorbance of this solution is measured at 518 nm.

50, 100, 200, 400 and 600 mg of the ester under determination is dissolved separately in 3ml of 0.25M NaOH. The volume in each is made up to 5 ml, the previously described procedure is followed. 5 ml of (i) is added as usual before recording absorbance of the individual tube for preparation of a calibration graph.

PROPIONIC ACID AND PROPIONATES

Propionic acid can be used for preserving foods with high pH because of its low dissociation constant. Propionates are widely used as antimicrobial preservatives. Na/Ca, propionates are widely employed for the prevention of mold attack in sliced bread and to some extent for preservation of cheese. Their higher concentration is needed for food preservation. Secondary propionates are effective against molds and *Bacillus mesentericus,* the bacteria responsible for rope formation in bread. Due to their weak inhibitory action against yeast, propionates poorly inhibit yeast leavening in bakery goods. In USA, up to 10% solutions of Na and Ca salts are employed for surface treatment of hard cheese against mold attack and are deemed GRAS. Though they are permitted *for paneer and chhana* in India but they are prohibited for processed cheese. In UK propionic acid, its Ca/Na salts are permitted in bread and other bakery products upto a maximum concentration of 3000 ppm on the basis of wheat flour.

Under Rule 55 of the PFA Rules 1955, propionic acid and its salts can be used as shown in Table 2.6.

Table 2.6

S. No.	*Article of food*	*Preservative*	*Max permissible limit (ppm)*
1.	Paneer or Chhana	Propionic Acid including esters or salt there of	2,000
2.	Breed and Biscuits		5,000
3.	Cheese sliced/cut/ shredded		3,000

Characteristics of Acid and its salts

Propionic acid is an oily liquid, slightly pungent having disagreeable rancid odour. It is miscible with water, soluble in alcohol, ether, $CHCl_3$. Sodium propionate has transparent crystals and granules. It is deliquescent in moist air and has neutral or slightly alkaline reaction to litmus. 1 gm dissolves in about 1ml water, in about 0.65 ml boiling water and in about 24 ml alcohol at 25°C. It is most active at acid pH.

Ca propionate occurs as mono-or trihydrate as powder or monoclinic crystals. It is soluble in water, slightly soluble in methanol and ethanol. It is practically insoluble in acetone and benzene. Volatile acids like acetic, propionic, butyric and valeric are steam distilled and the distillate is concentrated after neutralization, chromatographed, sprayed with methyl red and bromocresol and then visualised.

Reagents

(i) ***Developing Solvent:*** Acetone: tertiary-butanol: n-butanol: NH :: 2 : 1 : 1 : 1 is prepared afresh.

(ii) ***Spray Reagents:*** 200 mg each of methyl red and bromothymol blue is added to a mixture of 100 ml of formalin and 400 ml of alcohol. pH is adjusted to 5.2 with 0.1N NaOH.

(iii) ***Standard acid Solution:*** *I* ml each of acetic, propionic, butyric and valeric acid is pipetted into 100 ml volumetric flasks separately, the volume in each above marked flasks is made up with water. 1 ml of each of these stock solutions is pipetted into separate 25 ml beakers and 1 ml of each is pipetted into another beaker (mixture). Each solution is neutralised with 0.1 N NaOH using cresol red indicator and evaporated to dryness; charring is avoided. The dried residues are dissolved in 0.5 ml H_2O. These solutions are used for chromatography.

Procedure

20 gms of the well mixed sample is steam distilled and 200 ml distillate is collected which is immediately neutralized with 0.1 N NaOH using cresol red indicator. It is then evaporated to dryness and dissolved in 0.5 ml H_2O. 1-2 μl is spotted along with the standards on Whatman No. 1 paper which is air dried. This paper is clipped to glass rod and suspended in a tank with 50 ml mobile phase in a trough. As the mobile phase is heavy, 3 clips are used to hold the paper to the glass rod to prevent sagging. The chromatogram is developed to about 2.5 cm from the top of the paper, removed and air dried. It is then sprayed with the spray reagent (ii) uniformly. Faint yellow spots indicate presence of acids, whereas heavier blue spots are due to Na^+. To intensify spots, the paper is exposed to ammonia fumes. Entire paper immediately turns

green and acids generally appear as red spots. Since colour of acids is not stable therefore the spots are pencil marked soon after these are completely developed. Their R_f values determined whence their identification is made.

References

1. *Encyclopaedia of Chemical Technology,* Volume 3, IIIrd Edn., A Wiley-Interscience Publication, John Wiley and Sons, 1978, pp. 778.

Sodium metabisulphite

Also known as sodium pyrosulfite: $Na_2S_2O_5$: It crystallizes from cold water with $7H_2O$. It is available as white crystals or powder having odour of SO_2. Freely soluble in water, glycerol; slightly soluble in alcohol. The aqueous solution is acid.

Sodium sulphite

The marketed product is in the form of small crystals or a powder. It is fairly stable and does not oxidize as readily as hydrated sulphite. Soluble in 3.2 part water; in glycerol; almost insoluble in alcohol. pH about 9. It must be kept will closed.

Potassium metabisulphite

K pyrosulphite is its other name. Commercial product contains nearly 95% $K_2S_2O_5$. it is available either as white crystals as crystalline powder, has the odour of SO_2, is acidic in reaction; liberates SO_2 with acids; oxidizes in air to SO_4, more readily in the presence of it moisture. Catches fire if much heat develops while powdering. Freely soluble in water; insoluble in alcohol hence must be kept dry and well closed.

Though used as a preservative for fruits and vegetable, it is an anti-fermentative in breweries and wineries.

Chapter 3

High Intensity–Low Calorie Sweeteners

INTRODUCTION

Use of sugar in food items and drinks is as prevalent as use of salt. It is a necessary ingredient in the preparation of many items. Sweet tooth is common to both young and old. A strong will is required if sugar is to be avoided even if due to health reasons. It is indeed sweet on scientists part who have contributed towards reducing the miseries of having to do without sugar by developing substitutes-the present day intense sweeteners, i.e. the low-calorie sweeteners. While satisfying the desire for sweetness, they do not change appreciably the blood glucose level because they are usually non-carbohydrate substitutes with no food value but are useful as sugar substitutes in diets and beverages.

Naturally occurring non-sugar sweeteners are also worth considering because some of them are used traditionally. Development of new sweeteners require basic studies on the relationship of the chemical structure to the perception of sweet taste. Based on this understanding direct synthesis of newer classes of compounds as potential sweeteners needs to be undertaken. The Indian consumer should also be enlightened to the advantages of these scientific advances. Leading to newer, more acceptable and safer sweeteners which are intensely sweet, thereby lowering the quantities needed, are not metabolised hence are labelled as low-calorie sweeteners.

There are two broad categories of low-calorie sweeteners: (i) bulk sweeteners and (ii) high intensity sweeteners.

Low calorie bulk sweeteners include sugar alcohols such as sorbitol, mannitol, xylitol, erythritol and fructo-oligosaccharides such as insulin derivatives. Many of the bulk sweeteners are found in natural foods, but they can also be obtained industrially.

High intensity, low calorie sweeteners include saccharin, cyclamate, acesulfame-k, aspartame, thaumatin and sucralose. Most of them are synthetic compounds. All low-calorie sweeteners are not alike.

Bulk sweeteners differ from sucrose principally in being indigestible, but do contribute to the weight and volume of the food. These may get fermented by bacteria residing in the large intestine.

High intensity sweeteners are needed in very small quantities hence they add little other than sweetness to food. But they make food more palatable and palatability stimulates appetite leading to more intake of food without being hungry thus defeating the very purpose of low-calorie intake.

Low calorie sweeteners may not be completely inert. Besides the effect of palatability, one of the products of digestion of Aspertame is phenylalanine. Phenylalanine releases cholecystokinin, thought to induce satiety. But swallowing Aspertame in capsules has been variously reported to induce food intake or have no effect on it.

Sweet taste may itself stimulate the cephalic phase of insulin secretion, which can stimulate appetite. Drinking of sweetened solution of aspartame, saccharin or sucrose did not stimulate insulin secretion but sham feeding of apple pie did. Finally, *in vitro* and *in vivo* studies in rats have shown that acesulfame-k releases insulin but the dose required is higher than that which is required for use as a sweetener. Thus any effect or lack of effect of low-calorie sweeteners on food intake is consistent with the known physiological effects. Various studies have yielded conflicting results on human beings as psychologic and metabolic factors are side by side involved.

Nevertheless, from the available knowledge the following inferences have been drawn.

(i) Sweeteners by themselves neither promote weight gain/loss.
(ii) Low-calorie sweeteners achieve uncoupling of sensory and caloric characteristics and can therefore sweeten foods without adding to their caloric content

(iii) If the consumer uses this knowledge as an excuse for increasing food intake, the calorie-saving effect of the sweetener may be partly, fully or more than fully neutralized.

(iv) On the other hand, in subjects well motivated for losing weight, the possibility of reducing energy intake offered by the sweeteners is likely to be used profitably for reducing the hardship involved in realising their aim.

(v) Sucrose gets metabolised by bacteria in the oral cavity. The resulting acidic metabolites dissolve dental enamel making adverse effect on dental health. Aspartame, saccharin, ace sulfame-K, sorbitol and xylitol are all non acidogenic hence do not promote dental caries or plaque formation.

(vi) Intense sweeteners are not anorectic agents and will not suppress appetite. However, if an intense sweetener is used to replace sugar in an item, such as a drink, it is possible for the individual to use the calories saved to enjoy another product. Such an approach does not result in weight loss, but can help to prevent increase in weight.

INTENSE SWEETENERS-INDIAN SCENARIO

Clear understanding of the technical and economic aspects of intense sweeteners is a must in the Indian context. Production of intense sweeteners are fairly expensive and it is unlikely that unscrupulous small-scale producers will have any cost incentive in using them to replace sugar, a commodity of which India is the largest producer in the world and whose prices are highly controlled. In the wake of dereservation of biscuit and ice-cream production, large companies will enter this area and it is these large producers who shall have the money and technique to use intense sweeteners and it is the upper class consumers who will use their products.

Further these manufacturers will find it more profitable to develop a whole new range of low-calorie, low sugar products like their counterparts in the developed world to cater to the needs and desires of diabetics, slimmers and those with carcinogenic problems. The medical fraternity can be coordinated for the encouraged consumption of such products for health reasons.

All the detailed data on intense sweeteners is the result of the numerous and rigorous tests conducted by the FDA, the European

Union and the JECFA. The Indian scientists, technologists and regulatory authorities want to be sure that intense sweeteners do not behave differently in Indian condition and foods. Convincibility is lacking about the Indian physiological and nutritive aspects vis a vis the aforesaid studies. Neither is there an impetus to go in for these studies. But if undertaken and the use of intense sweeteners finds approval then the emerging scenario in India can be summed up as below:

(i) It shall be a boon for diabetics, dental patients, obese and weight watchers because the need for therapeutic low-calorie, low-sugar products exists but these patients have no choice but to use only Aspertame as a medicine and enjoy several items of food where sodium *saccharin* has been permitted as detailed elsewhere in this compendium.

(ii) Manufacture of intense sweeteners is not complicated and the capability of the Indian industry is much updated hence a vast scope for this industry can be harvested. *Saccharin* is being manufactured in India since long. Sucralose production facilities can make it available cheaper than the imported one because the basic starting material is sucrose and is available in abundance.

(iii) By providing an alternative to sugar, intense sweeteners shall spare tremendous domestic demand for sugar. This shall be of advantage to sugar exports where it can fetch much higher prices.

(iv) Indian food technologists and industries have certain reservations about the use of these intense sweeteners as a total substitute for sugar. They feel that intense sweeteners shall not be able to provide viscosity in syrups and fruit drinks, gelling action of *sucrose when* combined with a certain amount of hydrocolloids and calcium, preservative action in squashes where SO_2, and benzoates use depends upon the concentration of sucrose in them, natural flavour generation in bakery products, leavening action of sucrose in cakes and gas entrapment in ice creams and building action of sugar in many traditional Indian sweets like *gajak, rewadi,* etc.

Most data on the stability of intense sweeteners is based at 20°C which is too low for Indian conditions where ambient temperatures rise up to 50°C in some areas though 37°-40°C is normally used for testing thermostability. Studies on the thermostability of intense sweeteners in baking are very few, that too in the West. The studies reveal how well cyclamates, Aspertame and ace sulfame-k remain stable when one actually bakes the product; their inertness towards other food ingredients in Indian cooking also needs to be ascertained because Indian cooking is relished due to generation of flavours. While using intense sweeteners alone higher amylase activity may be imperative because glucose has to be generated *in situ* the likely flavouring element in the bread.

ARTIFICIAL SWEETENERS

Artificial sweetener is the legal term for "low-calorie sweeteners or intense sweeteners", however in the ensuing text these shall be referred as intense sweeteners though sometimes these are called "synthetic sweeteners" despite the fact that some of these are plant derivatives. These sweeteners have been in use since the discovery of saccharin in the 1880s but its widespread commercial application started only between forties and fifties, pushed by the sugar rationing introduced during second world war and by the efforts of saccharin's competitors to find an alternative. The real push came because the medical community increasingly sought sugar replacements that could be used in diabetic and dietetic treatments. Consequently, a tremendous boom occurred between seventies and eighties in the development and commercial use of sweeteners. New compounds were developed that were sweeter and more stable than their predecessors. Simultaneously, risk determination and safety testing methods underwent dramatic advances, enabling food product companies, food technologists and food regulation authorities to determine the specific nutritional and health effects of sweeteners in a variety of food products and a diversity of climatic conditions.

As public confidence in the developed world increased for intense sweeteners, use restrictions in national food laws began to be lifted leading to the use of these sweeteners, alone or in combination, in virtually all beverages and food products. This liberalisation has, in turn, blazed the development and widespread consumption of a vast selection of innovative, low calorie products and the rapid growth of new market niches within the food industry which caters to the needs of diabetics, obese/overweights and those with dental sensitivities.

Intense sweeteners are constituted from compounds that mimic the effects of sugar on the tongue, but do not react with the biochemistry of the human anatomy in the same way. In other words, they pass through the human body unmetabolised, i.e. without producing any calories. Aspertame is an exception, however, the body breaks it down into naturally occurring aminoacids and menthol and absorbs these as usual.

Each sweetener behaves differently in different applications hence has its own set of strengths and limitations. Rigorous food safety tests, often running up to 15 years are conducted on each intense sweetener under a variety of food product applications. When the intense sweetener has been found suitable, only then is it released for widespread commercial use.

Currently seven low-calorie intense sweeteners are available for commercial use in different overseas markets. But the most popular ones are saccharin, aspartame, acesulfame and sucralose—the latest addition and the only low-calorie intense sweetener actually manufactured from sugar. Cyclamate is also commonly used in Europe.

India's PFA. Act and its Rules appear to shut the above opportunities of the food industry and impose constraint on the use of these intense sweeteners. In other countries governments have radically rewritten or amended their national laws.

Rule 47 of PFA Rules 1955, stipulates that "no artificial sweetener shall be added to any article of food", provided, artificial sweeteners may be used in following food articles mentioned in Table 3.1 in quantities not exceeding the limits shown against them and shall bear the label declaration as provided in clause(1) and (2) of sub-rule (zzz) of Rule 42.

[1][**Restriction on use and sale of artificial sweeteners.**—[2][(1) No artificial sweetener shall be added to any article of food:

[3][Provided that artificial sweetener may be used in following food articles in quantities not exceeding the limits shown against them and as per provision contained in Appendix C to these rules and shall bear the label declarations as provided in sub-rule (ZZZ) (I), (ZZZ) -(1) (A), (ZZZ) (1) (B) and (ZZZ) (12) of rule 42.]

Notes

1. Subs, by G.S.R. 454(E), dated 15th April, 1988 (w.e.f. 15-4-1988).
2. Subs, by G.S.R. 284(E), dated 29th May, 1997 (w.e.f. 29-5-1997). Earlier it was subs, by G.S.8 695(E), dated 9th November, 1993 (w.e.f. 9-11-1993),
3. Subs, by G.S.R. 388 (E), dated 25th June, 2004, for the proviso (w.e.f. 25-6-2004). The provision before substitution, stood as under:-

Table 3.1 [1]

Sl. No.	*Name of artificial sweetener*	*Article of food*	*Maximum limit of artificial sweetener (ppm)*
1.	Saccharin Sodium	Carbonated Water	100
		Soft Drink Concentrate	100
	—do—	Supari	4000
		Pan Masala	8000
		Pan Flavouring Material	8.0%
		Synthetic Syrup for dispenser	450
		Sweets (Carbohoydrates based and Milk products based)—Halwa, Mysore Pak, Boondi Ladoo, Jalabi, Khoya Burfi, Peda, Gulab Jamun, Rasogolla and Similar milk product based sweets sold by any name.	500
		Chocolate (White, Milk, Plain, Composite and Filled)	500

1. Subs, by G.S.R. 388 (E), dated 25th June, 2004, for the Table (w.e.f. 25-6-2004).

Sl. No.	Name of artificial sweetener	Article of food	Maximum limit of artificial sweetener (ppm)
		Sugar based/Sugar free confectionery	3000
		Chewing gum/Bubble gum	3000
2.	Aspartame (methylester)	Carbonated water	700
		Soft Drink concentrate	700
		Biscuits, Bread, Cakes and Pastries	2200
		Sweets, (Carbohydrates based and milk products based)—Halwa, Mysore Pak, Boondi Ladoo, Jalabi, Khoya Burfi, Peda, Gulab Jamun, Rasogolla and Similar milk product based sweets sold by any name.	200
		Jam, Jellies, Marmalades	1000
		Chocolate (White, Milk, Plain, Composite and Filled)	2000
		Sugar based/Sugar free confectionery	10000
		Chewing gum/Bubble gum	10000
		Synthetic Syrup for dispenser	3000
3.	Acesulfame Potassium	Carbonated water	300
		Soft Drink concentrate	*300
		Biscuits, Bread, Cakes and Pastries	1000
		Sweets, (Carbohydrates based and milk	500

Sl. No.	*Name of artificial sweetener*	*Article of food*	*Maximum limit of artificial sweetener (ppm)*
		products based)—Halwa, Mysore Pak, Boondi Ladoo, Jalabi, Khoya Burfi, Peda, Gulab Jamun, Rasogolla and similar milk product based sweets sold by any name.	
		Chocolate (White, Milk, Plain, Composite and Filled)	500
		Sugar based/Sugar free confectionery	3500
		Chewing gum/Bubble gum	5000
		Synthetic Syrup for dispenser	1500
		Ready to serve tea and coffee based beverages	600
		Ice lollies/ice candy	800
		[3]Cereal based beverages	500
4.	Sucralose	Carbonated water	300
		Soft Drink Concentrate	300
		Biscuits, Bread, cakes and Pastries	750
		Sweets, (Carbohydrates based and Milk product based)—Halwa, Mysore Pak, Boondi Ladoo, Jalabi, Khoya Barfi, Peda, Gulab Juman, Rasogolla and similar milk product based sweets sold by name.	750

Sl. No.	*Name of artificial sweetener*	*Article of food*	*Maximum limit of artificial sweetener (ppm)*
		Ready to serve tea and coffee based beverages	600
		Ice lollies/ice candy	800
		Vegetable juice	250
		Vegetable nectar	250
		Concentrates for vegetable juice	1250
		Concentrate for vegetable nectar	1250

Explanation I—Pan flavouring material refers to the flavouring agents permitted for human consumption to be used for pan. It shall be labelled as—

PAN FLAVOURING MATERIAL

Explanation II.—Maximum limit of artificial sweetener in soft drink concentrate shall be as in reconstituted beverage or in final beverage for consumption. Soft Drink concentrate label shall give clear instruction for reconstitution of products: for making final beverage:

[2][Provided further that Saccharin Sodium or Aspartame (Methyl ester) or Acesulflame Potassium or sucralose] may be sold individually as Table Top Sweetener [3][and may contain the following carrier or filler articles with label declaration as provided in sub-clauses (1) and (2) of sub-rule (ZZZ) of rule 42, namely:—

1. Dextrose	8. Calcium silicate	15. Magnesium stearate IP
2. Lactose	9. Carboxymethyl Cellulose	16. Purified Talc
3. Maltodextrin	10. Cream of Tartar, IP	17. Poly vinyl pyrrolidone
4. Mannitol	11. Cross Carmellose sodium	18. Providone
5. Sucrose	12. Colloidal silicone dioxide	19. Sodium hydrogen carbonate
6. Isomalt	13. Glycine	20. Starch
7. Citric acid	14. L-leucine	21. Tartaric acid]

[4][Provided also that where sucralose is marketed as Table Top Sweetener, the concentration of sucralose shall not exceed six mg per hundred mg of tablet or granule.]

(2) No mixture of artificial sweeteners shall be added to any article of food or in the manufacture of table top sweeteners:

[5][Provided that in case of carbonated water, softdrink concentrate and synthetic syrup for dispenser wherein use of aspartame and acesulfame potassium have been allowed in the alternative, as per Table under- sub-rule(1), these artificial sweeteners may be used in combination with one or more alternative if the quantity of each artif ic·al sweetener so used does not exceed the maximum limit specified for that artificial sweetener in column (4) of the said Table as may be worked out on the basis of proportion in which such artificial sweeteners are combined. The products containing mixture of artificial sweeteners shall bear the label as provided under sub-rule (12) of sub-rule (ZZZ) of rule 42.]

Illustration.—In column (3) of the said Table, in carbonated water, Aspertame (Methyl Ester) or Acesulfame Potassium may be added in the proportion of 700 ppm or 300 ppm respectively. If both artificial sweetners are used in combination and the proportion of Aspertame (Methyl Ester) is 350 ppm, the proportion of Ascsulfame Potassium shall not exceed the proportion of 150 ppm.]

[(3) No person shall-sell table top sweetener except under label declaration as provided in clauses (1) and (2) of sub-rule *(ZZZ)* of rule 42:

Provided that Aspertame may be marketed as a table top sweetener in tabler or granular form in moisture-proof packages and the concentration of Aspertame shall not exceed 18 mg per 100 mg of tablet or granule.]

Rule 42(zzz)(2) stipulates that every package of Aspertame (methyl ester), acesulfame-K and saccharin sodium marketed as table top sweeteners and every package of carbonated water/ synthetic soft drink concentrate containing either of these artificial sweeteners and every advertisement for such table top sweetener or such carbonated water/synthetic soft drink concentrate shall carry the following label namely,

"Contains - - - - - - - - - - - (name of artificial sweetener)
Not recommended for children"

Provided the package of Aspertame (methyl ester) marketed as table top sweetener and every package of food containing Aspertame and the advertisement for such table top sweetener and food shall carry the following label namely,

"Not for Phyenlketoneuriscs"

A.07.10 of Appendix B stipulates following standards of pure saccharin sodium commonly known as soluble saccharin having empirical formula as-$C_7H_4NNaO_3$ S.$2H_2O$ and mol. wt. as 241.2 shall be the material which is soluble at 20°C in 1.5 parts of water and 50 parts of alcohol (95%), and shall contain not less than 98.0%/L and not more than the equivalent of 100.5% of $C_7H_4O_3$, NSNa calculated with reference to the substance dried to constant weight at 105°C, assay being carried as presented in I.P. It shall not contain more than 2 ppm of arsenic and 10 ppm of lead. The mt. pt. of saccharin isolated from the material as per I.P. method shall be between 226 °C and 230 °C. The loss on drying of the material at 105 °C shall not be less than 12.0% and not more than 16.0% of its weight. The material shall satisfy the tests of identification and shall conform to the limit tests for free acid/alkali, ammonium compounds and p-sulphamoylbenzoate as mentioned in the I.P.

A. 07.12 of Appendix B stipulates following standards of pure Aspartame, i.e. Aspartyl phenylalanine methyl ester having empirical formula as C_{14} H_{18} N_2O_5 and mol. wt. 294.31, shall be the material which is slightly soluble in water and methanol. It shall contain not less than 98 and not more than 102% of Aspertame on dried basis. It shall contain not more than 3 ppm of arsenic and 10 ppm of lead.

The loss on drying of the material at 105°C for 4 hours shall not be more than 4.3% of its weight. The sulphated ash shall not be more than 0.2%. It shall not contain more than 1% of diketo-piperzine.

A. 07.13: Acesulfame K:– Commonly known as such by this name has empirical formula $C_4H_4KNO_4S$, molecular weight as 201.24 shall be the material which is odourless, white crystalline powder having intensely sweet taste and is very slightly soluble in ethanol but freely soluble in water.

It shall contain not less than 99% and not more than 101% of Acesulfame—K on dried basis. It shall not contain more than 3 ppm fluoride. Heavy metals content shall not be more than 10 ppm. The loss on drying of material at 105 °C for 2 hours shall not be more than 1% of its weight.

MOLECULAR STRUCTURE AND SWEET TASTE

Historical

Early structure-taste theories attempted to identify certain common structural features of tastants and attribute a specific taste to these. One of the better known theories identified two distinct factors which should be present in a molecule for sweetness to be perceived, a glucophore (causing sweetness) and an otherwise tasteless entity, an auxogluc, which, when combined with the sweet glucophore, intensifies that sweetness. This was enunciated by Oertly *et. al.* They defined altogether six glucophores and nine auxoglucs. This theory could only be applied to explain the sweet tastes of aliphatic compounds and the sweetness of a number of intense sweeteners e.g. saccharin known at that time could not be explained.[1]

The importance of a migratory hydrogen atom for sweetness was proposed by Kodamma in 1920. This "vibratory" hydrogen could move from place to place, giving rise to different tautomeric forms. Loosely bound hydrogens are crucial to the present widely accepted theory of sweetness put forward by Shallenberger, *et. al.* This AH/B theory suggests that two H-bonds are formed between the tastant and the receptor. In the tastant molecule an electro negative centre (B) and a centre (HA) capable of donating a hydrogen atom are sought.[2]

Two complimentary sites are then required on the protein receptor so that the "Shallenberger Mechanism" can operate. The distance between AH to B should be approximately 3°A. The theory was satisfactory but other factors also had to be considered only then it could be explained that why all molecules possessing these sweet inducing entities were not sweet. Spatial, hydrophobic/ hydrophilic and electronic effects are very important and this becomes evident on examining some of the effects of molecular structure on taste for various classes of sweeteners. The strength of the Shallenberger-Acree theory and its extensions lie in its almost universal application. Other general theories of sweeteners have not been able to explain wide range like the AH/B theory hence this is regarded as the principle interclass theory of sweeteners. In the extension of AH/B theory, γ attempts to identify a third building site, variously labelled as the α, β or γ site, have been made, reflecting an interaction with the receptor through dispersion (hydrophobic) bonding by Kier[3], Shallenberger, *et. al*[4]. and Van der Heijden[5].

The optimum dimensions of the 'Kier triangle' are

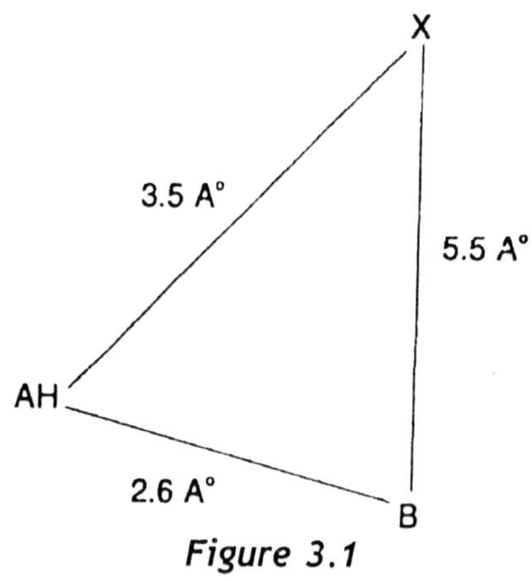

Figure 3.1

Classes of Sweeteners Studied for Molecular Structure-Sweetness Relationship (MS-SR)

1. ***Saccharin*** Unsurprisingly N-methyl saccharin is not sweet because the crucial H has been removed. The replacement of the CO by SO_2 group retains sweetness but when the reverse is done sweetness is destroyed.

Figure 3.2

2. ***Acesulfame-K*** Reaction of butyne and fluorosulfonyl isocyanate led to the synthesis of dimethyl compound (*) which was found to be sweet but subsequent synthesis of further dihydro oxathiazinone dioxides led to other compounds shown below including Acesulfame-K, the most suitable sweetener of this class for general use. The AH/B sites are indicated for ace-sulfame-K, but it is noteworthy that the B and X-sites will vary for the different derivatives shown.

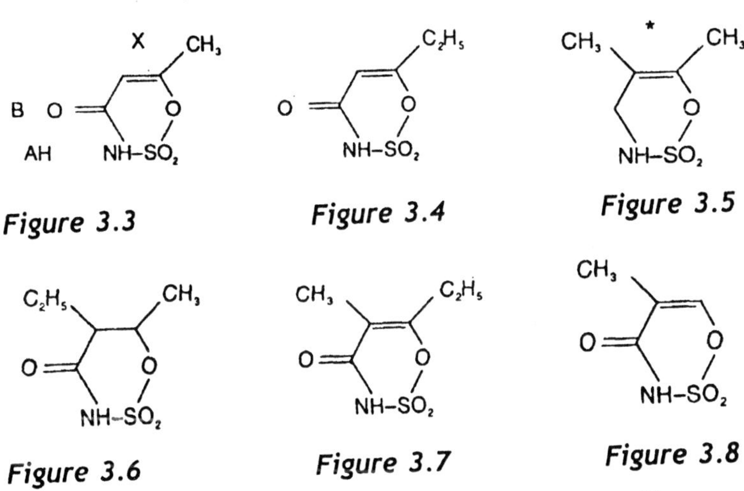

Figure 3.3

Figure 3.4

Figure 3.5

Figure 3.6

Figure 3.7

Figure 3.8

3. ***Nitroanilines*** The nitroaniline sweeteners were discovered in 1940s, including P-4000, the sweetest compound known to man (Fig. 3.9) in which R=O $(CH_2)_2 CH_3$. The carcinogenic properties of aromatic amines render them commercially useless. Nevertheless, these have been widely used in quantitative structure-activity relationship (QSAR) studies, since relative sweetness data are available for a fairly representative set. On such QSAR indicates that *n*, the hydrophobic parameter, and 5 the Wammet electronic

constant, show that hydrophilic and electronic effects are important in determining the relative sweetness of these compounds.

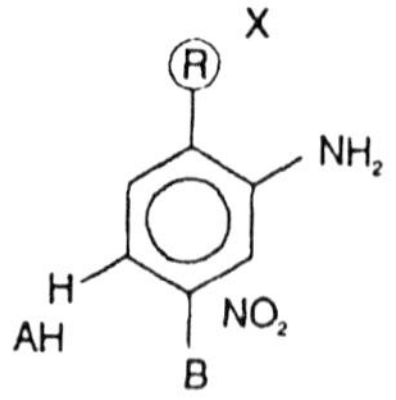

Figure 3.9

Table 3.2.: Modification of Sweetener Properties of Nitroanilines

Nitroanilines R	*Log RS**
$O\,(CH_2)_2\,CH_3$	3.613
$O\,CH_2\,CH = CH_2$	3.301
I	3.097
$O\,CH_2\,CH_3$	2.978
$O\,(CH_2)_3\,CH_3$	3.000
Br	2.903
$O\,CH\,(CH_3)_2$	2.778
Cl	2.602
OCH_3	2.342
CH_3	-
F	1.502
H	-

* $\log RS = 1.61\,\pi - 1.830\sigma + 1.73$, $R = 0.936$. $S=0.28$

4. ***Aspartame*** (L-Asparyl-L-Phenylalanine methyl ester) has the following configuration as shown in Fig. 3.10.

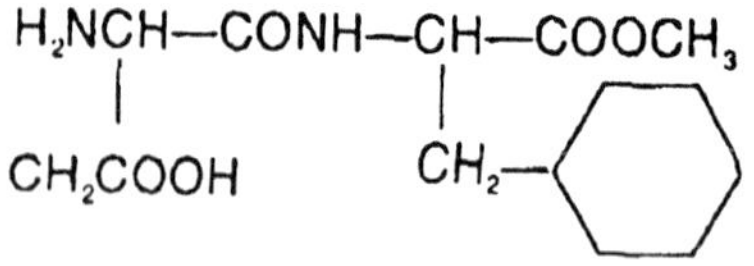

$H_2NCH—CONH—CH—COOCH_3$ with CH_2COOH on the first CH and CH_2—(ring) on the second CH

Figure 3.10

Figure 3.11 are the results of some structure-taste studies: Aspertame is the first entry and the last column gives the relative sweetness compared to sucrose. Though hundreds of derivatives/analogues of Aspertame have been synthesized and tasted, only a few results are given in Fig. 3.11.

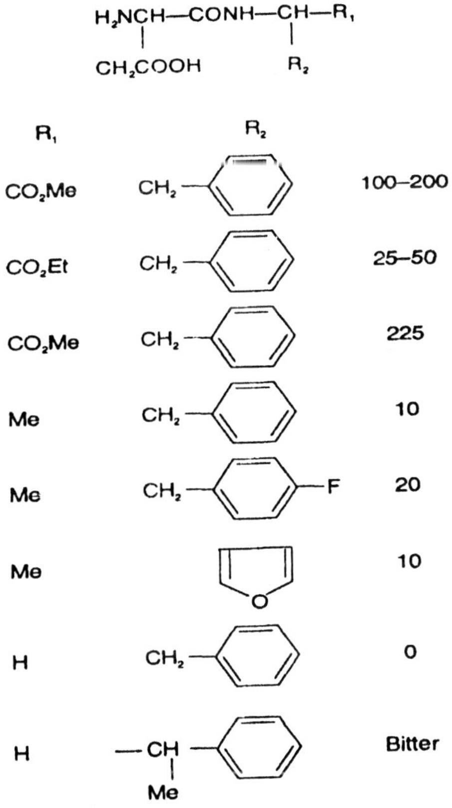

$$H_2NCH(CH_2COOH)—CONH—CH(R_2)—R_1$$

R_1	R_2	
CO_2Me	CH_2–C_6H_5	100–200
CO_2Et	CH_2–C_6H_5	25–50
CO_2Me	CH_2–C_6H_5	225
Me	CH_2–C_6H_5	10
Me	CH_2–C_6H_4–F	20
Me	furan ring (O)	10
H	CH_2–C_6H_5	0
H	–CH(Me)–C_6H_5	Bitter

Figure 3.11: Structure-taste studies related to aspertame.

5. ***Designer Sweeteners*** In a major development in the sweetener field, Tinti, *et. al.*[6] combined two known sweeteners, and named it Superaspartame, which is about 8000 times sweet as sucrose and much sweeter than the components from which it was made. Superaspartame combines not only the common 2-aminomelonyl function of Aspertame and cyanosuosan, but also other molecular features of the individual 'parent' sweeteners. These workers have identified besides AH/B/X sites additional D and G sites.

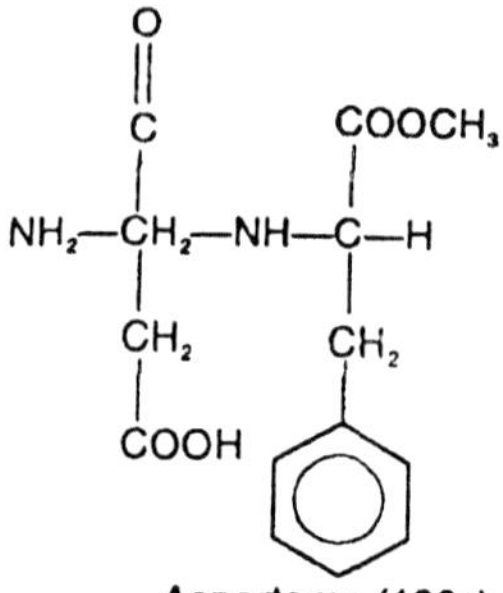

Aspartame (180x)

Figure 3.12

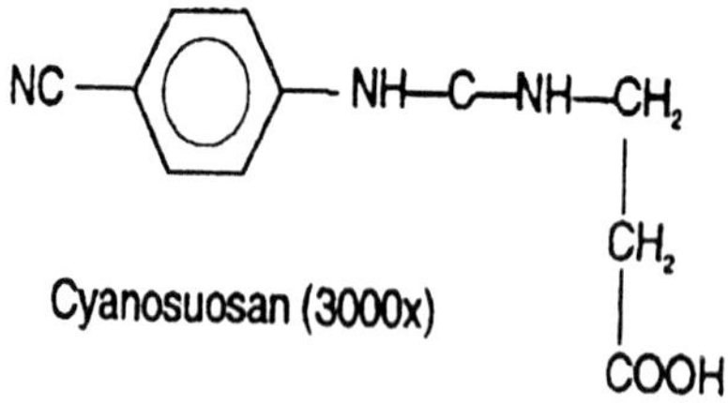

Cyanosuosan (3000x)

Figure 3.13

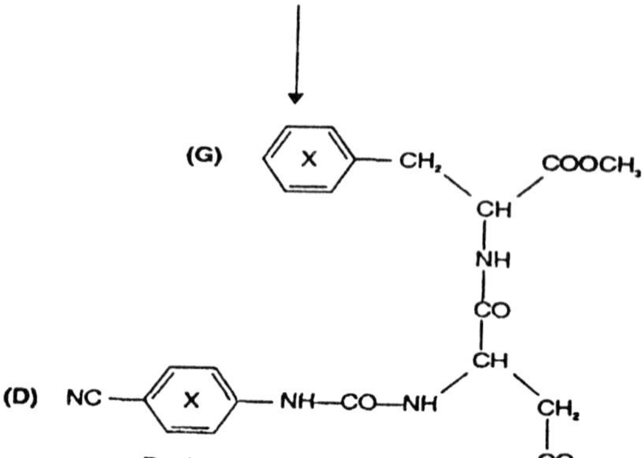

Design of Superaspartame (8000x)

Figure 3.14

Recent work led to the development of sucrononic acid, which is almost 200,000 times sweeter man sucrose. Its configuration is

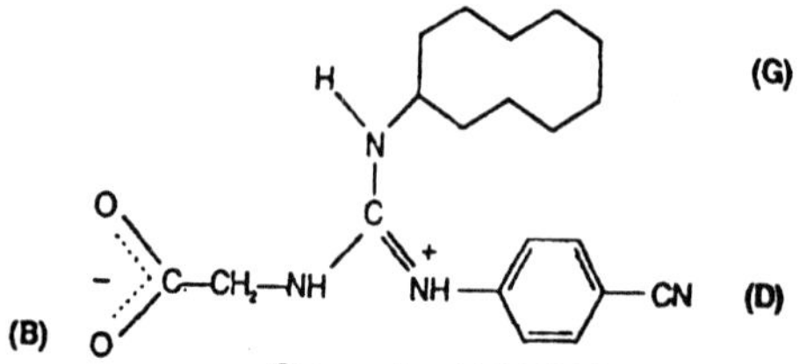

Sucronomic acid (20,000x)

figure 3.14

6. ***Naturally occurring sweeteners*** Monellin contains two amino acid subunits, and is about 300 times as sweet as sucrose. Thaumatin has a slight liquorice after-taste and is 750 times as sweet as sucrose. Micraculin is a taste modifier and causes

substances to taste sweet though it is not sweet itself. It is a glycoprotein and is deemed a miracle fruit. Circulin, from fruits of *Curculigo latifolia,* is a polypeptide, is sweet and also a taste modifier.

Stevioside makes up 6% by weight of the leaves of the Paraguayan plant *Stevia rebaudiana* and is about 300 times as sweet as sucrose. Osladin is a steriodal saponin about 3000 times as sweet as sucrose. Glycyrrhizic acid is about 50 times as sweet as sucrose. Periandrin, a terpene glycoside, has been isolated from the roots of the Japanese plant, *Periandra dulcis.* Its relative sweetness is similar to that of glycerrhizin.

7. *Ureas* Dulcin (about 250 times as sweet as sucrose) is the best known of the urea sweeteners. Sketches below illustrate the limitations in the positioning of the $-OC_2H_5$ and $-CH_3$ groups on the aromatic ring if sweetness is to be retained. Clearly shorter AH-X and B-X distances are not favourable for the retention of sweetness.

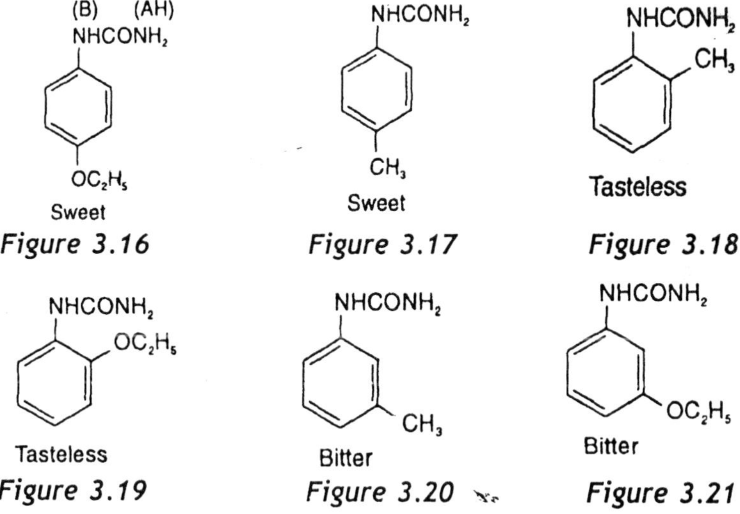

Figure 3.16 Figure 3.17 Figure 3.18

Figure 3.19 Figure 3.20 Figure 3.21

8. *Oximes* Perillartine (perillaldehyde oxime) is the purified oil from the Japanese plant *Perilla namkinensis.* Of the numerous synthesized/prepared compounds by Acton, *et.al.* are :–

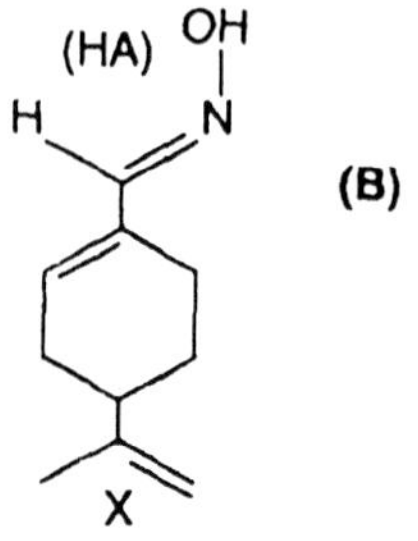

(Parent Compound)

Figure 3.22

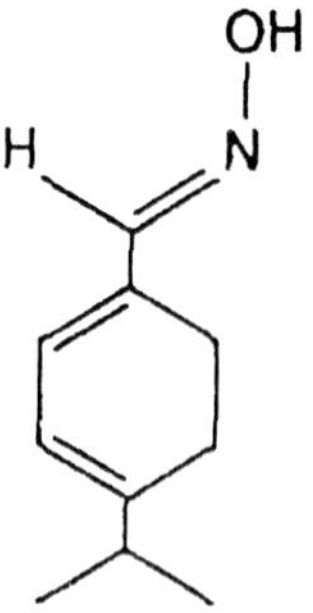

Figure 3.23

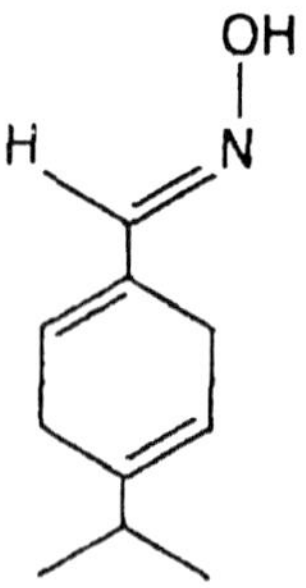

Figure 3.24

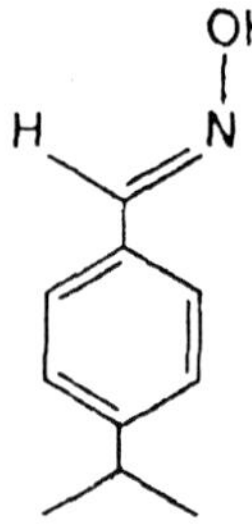

Figure 3.25

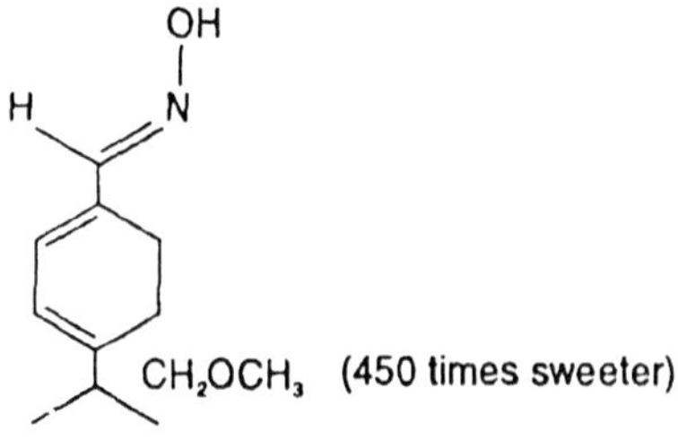

CH_2OCH_3 (450 times sweeter)

SRI Oxine–V

Figure 3.26

SRI oxime V overcomes the low water solubility of the other oximes, which restricts their usage A C = C conjugated to the oxime is an *essential requirement for sweetness.*

9. ***Isocoumarins*** Phyllodulcin, isolated from the leaves of *Hydrangea macrophylla (Seringe),* is about 200-300 times as sweet as sucrose. It is a 3, 4-dihydroisocoumarin derivative. Two AH/B locations are easily recognisable in these structures. A few structure-taste data showing modification of phyllodulcin sweetness is shown in Figs. 3.27-3.30.

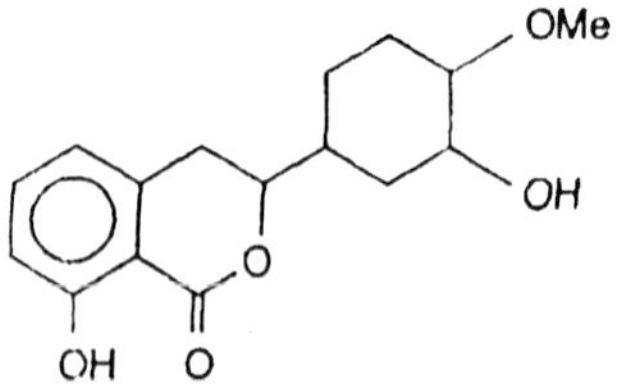

[Phyllodulcin (400x)]

Figure 3.27

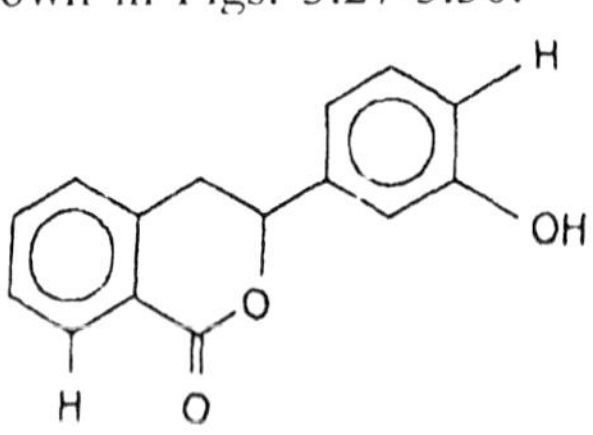

(Tasteless)

Figure 3.28

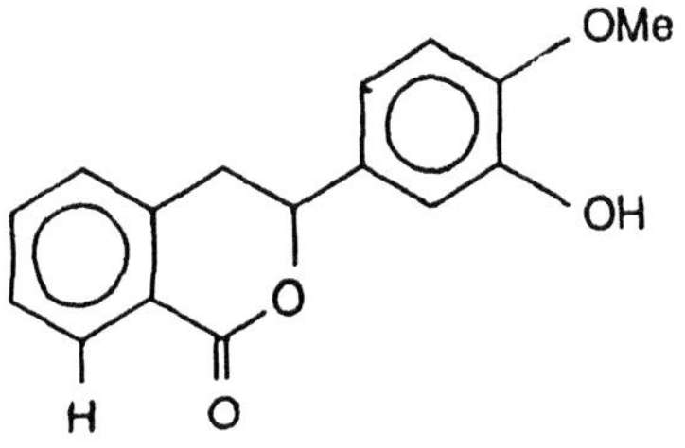

OMe OH H O [Sweet (100x)]

OH H OH O (Tasteless)

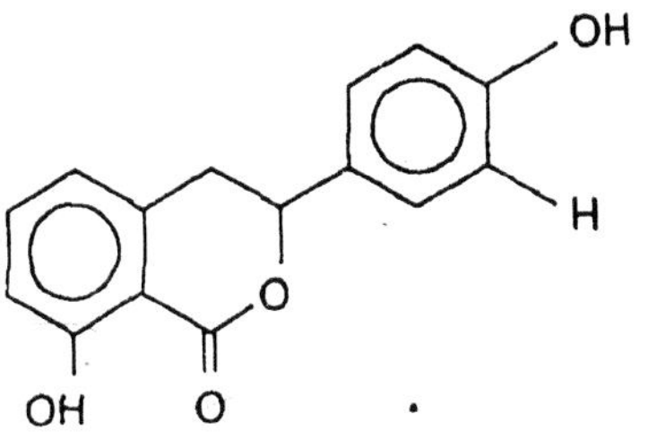

Figure 3.29 **Figure 3.30**

10. ***Flavones*** Structurally akin to phyllodulcins and are widely distributed in nature and many are sweet.
11. ***Dihydrochalcones (DHC)*** The parent compound is dihydrochalcone (sweet) and others are its substituted derivatives.

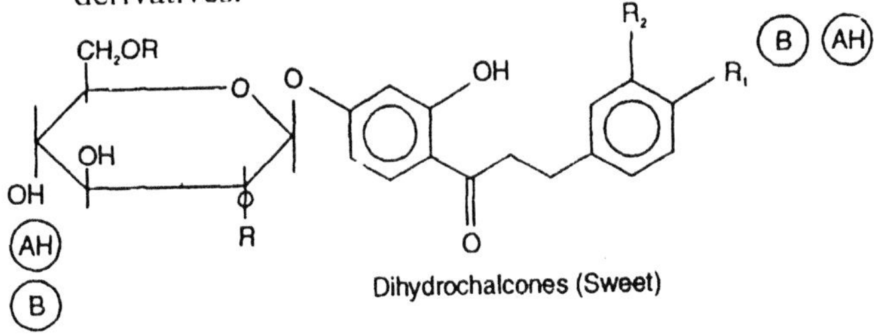

CH_2OR O O OH R_2 R_1 (B) (AH) OH OH (AH) (B) R O

Dihydrochalcones (Sweet)

Figure 3.31

1. Naringin DHC; R_1 - OH; R_2 = H; R = α - L - Rhamnosyl
2. Neohesperidin DHC; R_1 = OCH_3; R_2 = OH; R = α - L - Rhamnosyl
3. Hesperidin DHC; R_1 = OCH_3; R_2 - OH; R = H

The DHC sweeteners are prepared from naturally occurring bit ter tasting flavone from citrus fruits, and are thus examples of semi synthetic sweeteners.The three best knownDHC(s) are shown above and are prepared by chemical reaction from naringin, neohesperidin and hesperetin. They are nearly 200 times as sweet as sucrose and have a menthol like after-taste.

12. ***Isovanittyls*** AH/B sites indicated, along with their sweetening potential.

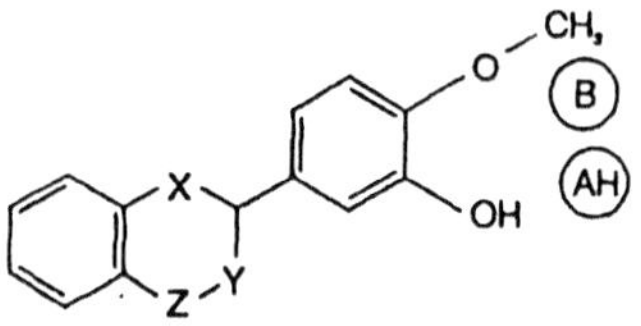

Figure 3.32

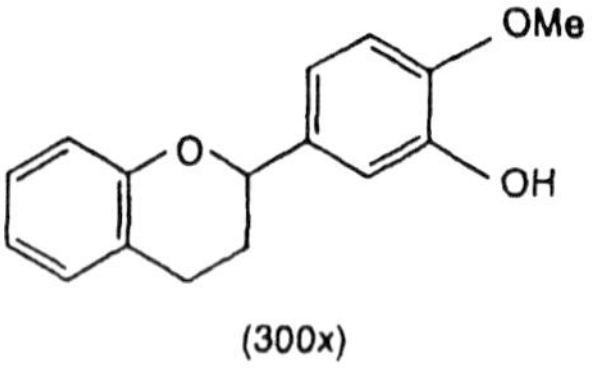

(300x)

Figure 3.33

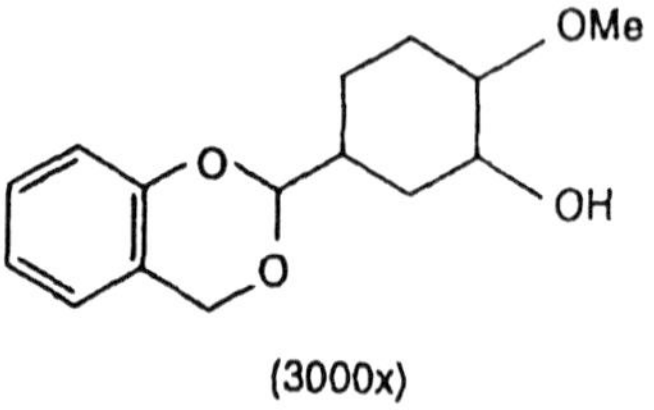

(3000x)

Figure 3.34

(450x)

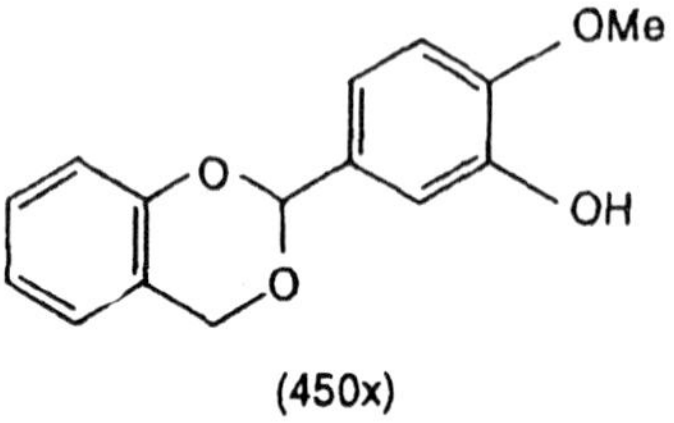

Figure 3.35

13. ***Sulfamates*** The parent compound is N-cyclohexylsulfamate (refer Fig. 3.36). The cation does not affect the relative sweetness (RS) which is about 40 (based on a 3% sucrose solution). Replacement of the amino hydrogen results in loss of total sweetness.

Severcl groups have been active in structure-taste studies. The following configurations show the effect of ring size, (refer to Figs. 3.37 (a, b, c) and 3.38 which show all sweet and Figs. 3.39-3.41 which show all non-sweet).

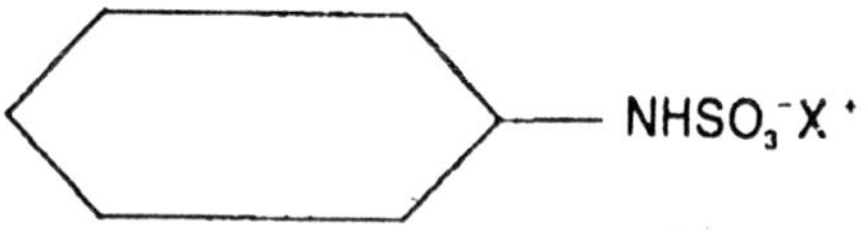

$NHSO_3^-X^+$

X = Ag, Na, NH_4, Cyc–$C_6H_{11}NH_3$

Figure 3.36

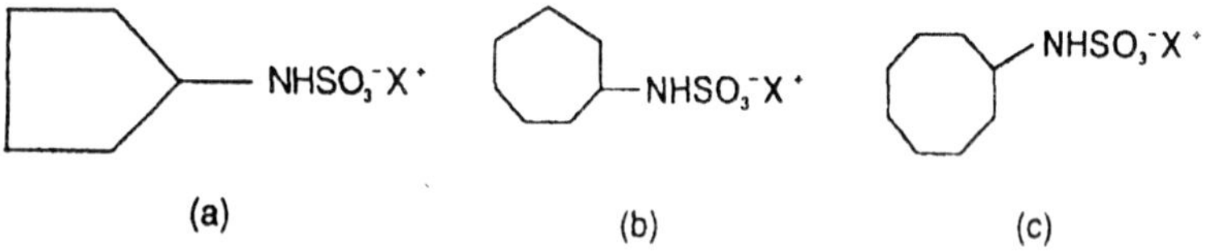

(a) (b) (c)

Figure 3.37

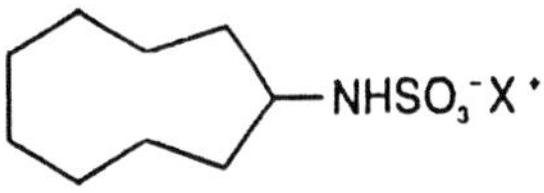

figure 3.38

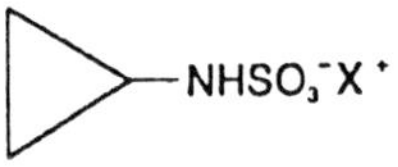

Figure 3.39 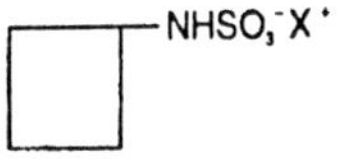 *Figure 3.40* 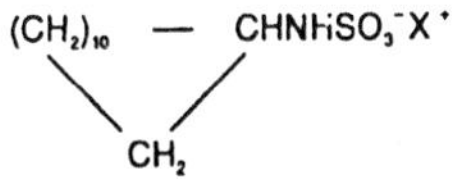 *Figure 3.41*

Present Research Position

Attempts are currently being made to develop a QSAR for about 20 structurally diverse sulfamates for which RS (relative sweetness) data exist, in the Department of Chemistry, University College, Galway. This is expected to be on the lines of the nitro-anilines. An extended sulfamation programme is being conducted from a diverse assay of compounds, (including nitroaniline and phenyl ura tastants) and synthesizing their mono and disulfamates. Some representatitives sulfamates.[8] are

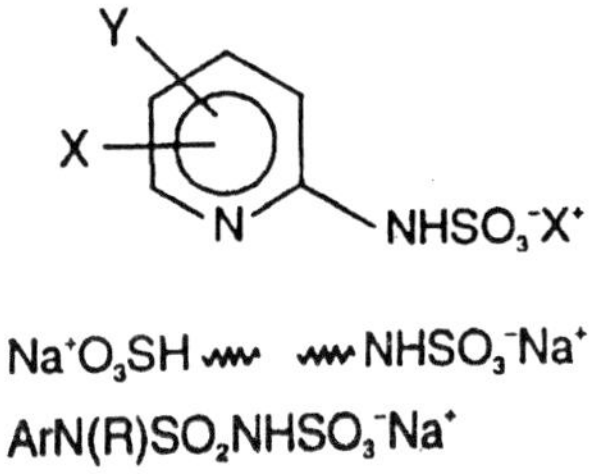

Figure 3.42 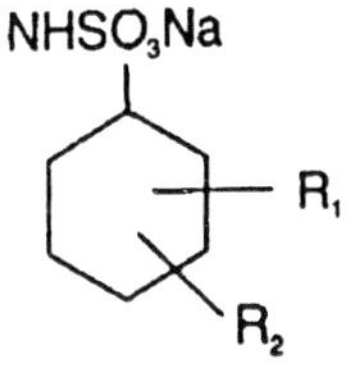 *Figure 3.43*

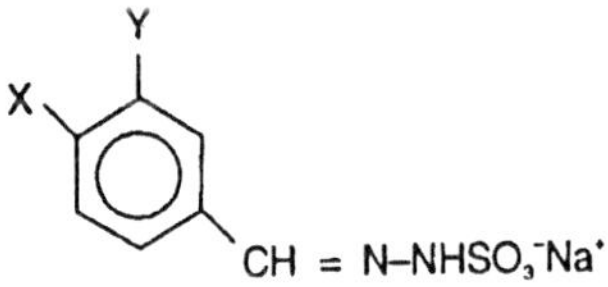

Figure 3.44

SACCHARIN

Saccharin is also called insoluble saccharin/benzosulfimide/o⁻ sulfo-benzimide, benzoic sulfimide, gluside, glucid, garantose, saccharinol, saccharinose, saccharol, saxin, sykose, hermesetas, etc. Its molecular formula is $C_7 H_5 NO_3 S$, mol. wt. is 183.18 and configuration is,

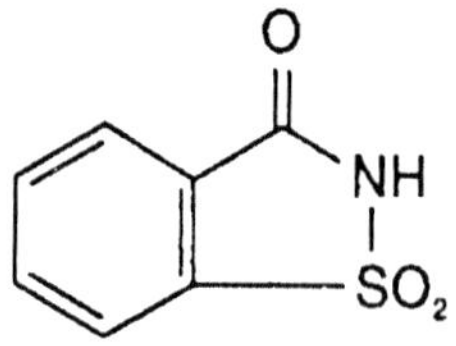

Figure 3.45

Monoclinic crystals having mt. pt 228-229.7°C, twinning on (001), perfect 100 cleavage make saccharin. Acicular crystals are prepared by vacuum sublimation. In dilute aqueous solution it is 500 times as sweet as sugar; the sweet taste is still detectable in 1:100,000 dilution. It has bitter metallic after taste, its density is 0.828, heat of combustion at constant volume is 4753.1 cal/g, UV_{max} (0.1 N NaOH) is 267.3 nm.

1 g dissolves in 290 ml water, 25 ml boiling water, 31ml alcohol, 12ml acetone and about 50 ml glycerol. It is freely soluble in solutions of alkali carbonates, slightly soluble in chloroform, ether. pH of 0.35% aqueous solution is 2.0. Alkaline hydrolysis yield O'-sulfamoyl benzoic acid and acid hydrolysis yields NH_4-O-sulfobenzoic acid.

Sodium salt dihydrate is more known by such names as *soluble saccharin, sodium saccharin, Kristallose, Crystattose, Dagutan, Sucaryl, Sucromat.* Crystalline powder effloresces in dry air. In dilute aqueous solution it is 300-500 times as sweet as sugar. 1 gm dissolves in 1.2 ml water and about 50 ml alcohol. Aqueous solutions are neutral or alkaline to litmus, but not alkaline to phenolapthalein.

Till 1985 it was considered to be carcinogen[9]. It was believed to have induced bladder cancer in male laboratory rats. Only in the nineties it has been confirmed that saccharin does not cause cancer in humans and other animals and is carcinogenic only in male rats when administered in extremely high doses because it reacts with phosphate, calcium and protein levels in the animal urine.

Over 90 countries have approved the use of saccharin. As stated earlier, in India saccharin is one of the four sweeteners presently permitted under PFA. Act 1954. It has been available in India for forty years and is the only low-calorie sweetener manufactured indigenously.

Saccharin was first produced in India in 1950. Till 1997, the installed capacity in Western India was 1500 tonnes/annum. Each and every raw material used in saccharin production is manufactured in India, hence it is a totally indigenous product.

Detection

Preparation of Sample

(a) ***Non-Alcoholic Beverages*** 25 ml of sample and 3 ml HCl are taken in a separating funnel. If vanillin is present it is removed by multiple extractions with pet ether, drained and discarded. $(C_2H_5)_2O$: pet ether :: 1 : 1 is used for extraction with 50, 25 and 25 ml portions. The combined extracts are washed with 5 ml of water and the solvent is completely evaporated.

(b) ***Semi-solid Preparations*** 25 g of the sample is transferred to a 100 ml volumetric flask with small amount of water. Boiling water is added to make up 75 ml volume then kept for one hour with occasional shaking. 3 ml of acetic acid is added and the contents are thoroughly mixed. Neutral lead acetate is added in slight excess and the volume is made up. The contents are again mixed and kept aside for 20 minutes and filtered. 50 ml of filtrate is taken in a separating funnel.

Detection Procedure

(i) ***By conversion to salicylic acid*** If salicyclic acid is not present in the sample the residue is dissolved in about 10 ml of hot water and 2 ml of dil. H_2SO_4 (1 : 3) is added. The contents are boiled, a slight excess of 5% $KMnO_4$ solution is added drop-wise and the solution is partly cooled. 1 g NaOH is dissolved in this solution and the mixture filtered into a silver crucible. The filtrate is evaporated to dryness then heated for 20 minutes at 213 ± 2°C. The residue is dissolved in hot water, acidified with HCl, a few drops of neutral 0.5% $FeCl_3$ solution are added. Violet colour indicates salicylic acid formation from saccharin.

(ii) ***Phenosulphuric Acid Test*** To the residue obtained after removing the solvent, 5 ml of phenol-H_2SO_4 reagent, prepared by dissolving pure colourless phenol crystals in equal weight of H_2SO_4, is added. The contents are heated for 2 hours at 138 ± 2°C, then dissolved in a little hot water, and made

alkaline with 10% NaOH. A magenta or reddish-purple colour develops if saccharin is present.

(iii) ***Resorcinol-H_2SO_4 test*** A 1:1 solution of resorcinol: H_2SO_4 is prepared and 5 drops of it are added to the residue. The reactant is heated on a low flame till a red colour develops. It is dissolved in 10 ml of water, made alkaline with 10% NaOH solution then few drops of iodine solution are added. A green fluorescence indicates presence of saccharin.

Estimation

Non-Alcoholic Beverages

Saccharin is extracted from a known quantity of the acidified sample with diethyl ether. After desolventisation, the residue is digested with HCl and diluted to a known volume. An aliquot is treated with Nessler's reagent and the absorbance of the coloured product is measured at 425 nm.

Reagents

(i) ***HCl***

(ii) ***Diethyl ether $(C_2H_5)_2O$***

(iii) ***Nessler's reagent*** 100 g of HgI_2 and 70g KI are dissolved in little water then slowly added to the cooled solution of 160 g NaOH in 500 ml water and diluted to litre.

(iv) ***Standard solution*** 0.2921 NH_4Cl is dissolved in 1 litre of ammonia free water = 1 g of saccharin/litre. It is diluted to have a 200 ng/ml equivalent of saccharin.

Procedure

50 gm of sample is accurately weighed in a separating funnel, 2 ml HCl is added, the contents mixed and extracted with 3 x 50 ml of (ii). Ether extract is filtered through cotton into a clean 250 ml conical flask and the solvent is evaporated. 6 ml HCl and 5 ml water is added, evaporated on a hot-water bath to 1 ml. This is diluted to 550 ml with NH_3-free water. In a 25 ml volumetric flask to 2 ml of this solution 1 ml (iii), is added the volume is made up. 0.5, 1, 2, 3 and 4 ml portions of (iv) (200 ug/ml) are taken in individual 25 ml volumetric flasks and the colour is developed with (iii). Product absorbance is read at 425 nm against similarly prepared reagent blank. Saccharin content is computed from the' calibration graph.

Phenol-H_2SO_4 Colourimetric Method

Saccharin is extracted from the acidified sample with $CHCl_3$ and benzene and the solvent evaporated. The residue obtained is treated with phenol-H_2SO_4 and heated at 175°C for 2 hours. The contents are made alkaline with NaOH and the absorbance is read at 558 nm.

Reagents

(a) $CHCl_3$
(b) $(C_2H_5)_2O$
(c) $CHCl_3$:benzene::95:5
(d) CH_3OH
(e) Phenol (colourless crystals)
(f) H_2SO_4
(g) Saccharin standard

1. ***Soft drinks*** Usually carbonated beverages and low-calorie drinks fall in this class. The former is decarbonated by repeated shaking and pouring back and forth in two beakers. 10 ml sample is transferred to a 125 ml separating funnel fitted with a Teflon stopcock. 15 ml water and 0.5 ml of IN NaOH are added, mixed and extracted with 50 ml of (c) after 1 minute shaking. The separated solvent layer is discarded, (benzoic acid and benzoates do not interfere).
2. ***Fruit juices*** 50 ml sample is taken in a 100 ml volumetric flask, a slight excess of 5% neutral lead acetate solution, not more than 10 ml is added, the volume is diluted with water, kept aside for 1 hour then filtered. A 2 ± 1 mg saccharin containing aliquot from filtrate is proceeded as described in the following section *Determination* from Step 1 onwards.
3. ***Food sweetener, tablets and concentrated liquids*** 15 ± 5 tablets are uniformly powdered. From it 500 mg is weighed accurately or 10 ml liquid concentrate transferred into a 500 ml volumetric flask, the volume is made up with water. 10-15 ml aliquot is taken for determination. If liquid concentrate contains parabens as preservatives then the aliquot is acidified with 5 ml dil HCl (1:4) and extracted with 20 ml CCl_4. The CCl_4 separated layer is discarded and proceeded from Step 2 onwards in the following section *Determination*.

4. ***Jellies and preservatives*** From the blended sample 25 g are taken into a 50 ml beaker which is heated on a water bath to liquify it. It is then transferred completely to a 250 ml volumetric flask using 25 ml hot water. The volume is made up with methanol and contents mixed thoroughly and left for 1 minute and finally filtered. An aliquot containing 2 ± 1 mg saccharin is transferred to a 50 ml beaker, evaporated to half the volume on water bath to remove alcohol and transferred to a 125 ml separating funnel with the help of about 25 ml hot water. After this follow from Step 1 onwards in the section *Determination.*
5. ***Low-calorie, high protein powders/granules/liquids*** Granules are uniformly powdered. 15 ± 5 g of powder is accurately weighed and transferred completely into a 250 ml volumetric flask containing about 150 ml hot water. Contents are homogenised, a slight excess of 5% neutral lead acetate solution (30 ml) is added and diluted to volume with cold water, mixed and kept aside for 1 hour to homogenise. For liquids 50 g sample is used and proceeded from Step 1 in the section *Determination.*
6. ***Chocolate Bars*** 25 g is weighed from the shredded sample in a beaker, 150 ml hot water is added and mixed with magnetic stirrer to disperse or emulsify. A slight excess of 50% neutral lead acetate solution (30 ml) is added to the contents and transferred to a 250 ml volumetric flask with water and diluted to volume, mixed and left for 1 hour and is then filtered. 50 ml filtrate aliquot is proceeded from Step 1 of the section *Determination.*

Determination

Aliquot as specified for the prepared sample is transferred along with (2 ± 1) mg saccharin) solution to the separating funnels.

Step 1 5 ml of dil. HCl (1:4) is added.

Step 2 It is extracted by shaking for one minute, each time with 50, 30 and 20 ml of (c) or with ether : benzene :: 95 : 5 as specified in sample preparation.

Step 3 The combined solvent extract is filtered through funnel fitted with pledget of glass wool and containing 10 gms anhydrous Na_2SO_4 into a 100 ml volumetric flask.

Step 4 The volume is made up with the same solvent mixture used and the contents in the flask are mixed.

Step 5 20 ml aliquot is transferred to a 50 ml Erlenmeyer flask.

Step 6 The solvent is evaporated to dryness in a shallow water bath and complete drying is accomplished in an oven at 100°C for 20 minutes.

Step 7 3 ± 2 ml hot melted phenol is pipetted to Erlenmeyer flask and the residue is dissolved by swirling the flask. 1.2 ml H_2SO_4 is pipetted and the contents reswirled.

Step 8 A Blank is prepared by pipetting 2.5 ml hot melted phenol and 1.5 ml H_2SO_4 into a 50 ml Erlenmeyer flask.

Step 9 The flask is stoppered with a tight cap covered with aluminium foil and heated for 2 hours at 175°C in an oven, cooled and approx. 20 ml of hot water is added to the flask and mixed.

Step 10 10 ml of 20% NaOH solution is added, mixed, quantitatively transferred to a 900 ml volumetric flask and diluted to volume with water.

Step 11 Absorbance of the solution is read in a spectrophotometer at 558 nm.

Step 12 Concentration of saccharin is determined by comparing with a calibration curve.

ASPERTAME

Aspertame also called Canderel, Equal, Nutra sweet, Sanecta, Tri sweet is chemically N-L-a-Aspartyl-L-phenylalanine 1-methyl ester. Its molecular formula is $C_{14}H_{18}N_2O_5$ and mol. wt. is 294.30. Its dipeptide ester is about 160 times sweeter than sucrose in aqueous solution.

Aspertame is unique among high intensity sweeteners because, immediately upon ingestion it is completely metabolised by enzymes in the gastrointestinal tract to three naturally occurring dietary components which are, aspartic acid, phenylalanine (constituents of normal protein) and the methyl ester forms methanol, like methyl esters in fruits and vegetables. Aspertame provides very small amounts of these components when compared to the amounts present in the daily diet (some where in the range 2-3% for heavy users)

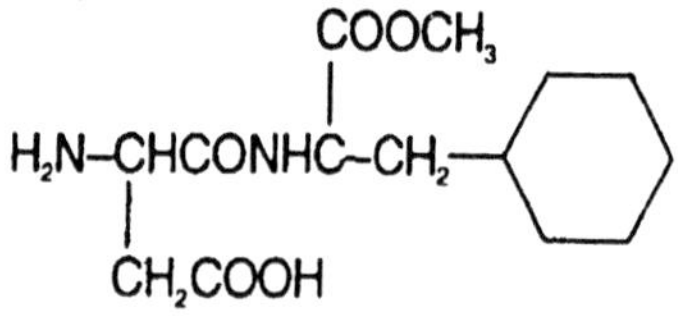

$COOCH_3$

$H_2N-CHCONHC-CH_2-$

CH_2COOH

Colourless needles from H_2O
mt. pt. 246–247 °C
$[\alpha]^{22}_{D}$ 2–3° (INHCl)

Figure 3.46

Aspertame is 200 times sweeter than sucrose, its 1 gram provides only 4 calories versus 800 for sucrose. When blended with acesulfame-K, the duo offers interesting possibilities to modify time intensity profiles. This is due to the sweetness of the latter perceived quickly, whereas aspertame's sweetness shows a slightly delayed onset but a more lasting taste. By choosing different blend ratios, stress can be put on the initial sweetness or a most lasting effect as desired. This modification of perception of sweetness is not possible with single sweeteners or several other sweetener blends.

Detection and Estimation (TLC) Aspertame is extracted from dry beverage powders using methanol: acetic acid :: 80 : 20 and isolated by TLC.

Reagents

(i) ***Methanol: acetic acid:: 80:20.***

(ii) ***Aspertame stock solution*** 50 mg of standard Aspertame is dissolved in about 80 ml of (i) and the volume made up to 100 ml in a volumetric flask.

(iii) ***Starch solution*** 600 mg of soluble starch is dissolved in 120 ml water, boiled for 10 minutes and filtered through medium filter paper (Whatman No. 2 or equivalent)

(iv) ***KI: Starch solution*** 500 mg of KI is dissolved in 100 ml of (iii)

(v) ***Developing solvent*** Methanol : glacial acetic acid : H_2O : $CHCl_3$:: 15 : 1 : 3 : 32.

(vi) ***Tertiary butyl hypochlorite (TBHC) chamber*** 5 ml of TBHC is added to a small glass vessel which can be capped and contains a glass wool wick reaching its top. This vessel is then uncapped in the TBHC chamber-15 minutes before use.

(vii) ***Silica Gel G***

Preparation of sample

(i) *Dry powders* Aspertame contained in the sample (dry powder) is extracted by shaking the sample with 100 ml of (i) for 20 minutes The quantity of the sample to be taken depends upon its theoretical Aspertame content. Assuming 100% extraction, 5g of the sample containing 0.4% aspertame, extracted into 100 ml of (i) would produce 2 μg/μl of supernatant. Since 2 μl are to be spotted, the resulting chromatogram should match the 4 ng standard spot which is achieved by spotting 2 ul of 2 μg/ml standard.

(ii) *Liquids* These are treated as above except that the extract is made up to 100 ml with (i).

Procedure

2 ul of the sample solution and standard using a micropipette is spotted and dried by a cool air stream. The plate is placed in the development chamber and the solvent is allowed to ascend about 15 cm. After removing the plate it is air dried for 15 minutes then placed in (vi) for 15 minutes and air-dried in fume hood for 30 minutes, sprayed with (iv). Aspertame is estimated by comparison with the standard spots.

Note: TBHC is toxic when absorbed through the skin or inhaled, therefore gloves and mask should be used while handling it.

A. UV spectrophotometric method Aspertame is extracted with aqueous methanol and the absorbance of the filtered solution is measured.

Reagents

(i) Solvent mixture 350 ml of water and 150 ml of methanol are mixed and allowed to equilibrate at room temperature.

(ii) Standard Aspertame solution 72 mg of Aspertame is dried at 105°C for 2 hours and is then transferred into a 100 ml volumetric flask, 50 ml of (i) is added and dissolved by shaking and the volume made up with (i).

Preparation of sample Since under Rule 47 (3) of PFA Rules (1955), aspertame content of tablet/granular package must be less than 18 mg/100 mg of tablet or granule package, therefore powdered tablet material equivalent to °4 tablets is accurately weighed and quantitatively transferred into a volumetric flask, 50 ml of (i) is added, shaken for 30 minutes on a flask shaker and volume made up with (i). Contents are filtered through Whatman No.1 filter paper, first 20 ml of filtrate is discarded and the rest is collected in a stoppered flask.

Procedure Absorbance of (ii) and test solution is measured at 258nm against (i) as blank/control. Aspertame content is calculated from these absorbances.

B. ***Non-Aqueous titration method*** Aspertame from the sample is extracted with dimethyl formamide and the solution is titrated with standard lithium or sodium methoxide using thymol blue as indicator. Aspertame content is calculated as usual in volumetric calculations.

Reagents

(i) ***Li/Na methoxide solution (0.01N)*** Prepared and standardised against standard benzoic acid using thymol blue indicator.

(ii) ***N-N-dimethyl formamide (DMF).***

(Hi) ***Thymol blue indicator*** 0.3% solution in (ii).

Procedure Accurately weighed, tablet or granule equivalent to 10 tablets is transferred to a 250 ml dry beaker then 50 ml of DMF is added and contents are warmed on a boiling-water bath, cooled and filtered through a sintered glass funnel into a 250 ml dry conical flask. The beaker and funnel are thoroughly rinsed with 50 ml DMF. Rinsings are also collected in the same conical flask. The homogenised contents are titrated against (i) using 4 drops of (iii) to a dark blue end point (T). A blank (B) is also done.

Calculation

Per cent of aspertame = 16.35 x (T - B) x N x 10, where N is normality of (i)

CYCLAMATE

Cyclamic acid, also called hexamic acid or cyclohexylsulfonic acid has $C_6H_{13}NO_3S$ as its molecular formula. Its molecular wt. is 179.24. It has the following structure.

The acid crystals are sour-sweet, having mt. pt. 169-170°C. It is fairly strong acid, very sparingly soluble in water and is slowly hydrolysed by hot water.

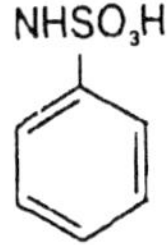

Figure 3.47

For these reasons, cyclamic acid itself is not used but its salts are widely used. The sodium salt, popularly known as sodium cyclamate, Assugrin, Sucaryl sodium or Sucrosa consists of pleasant, sweet to the taste crystals. It is freely soluble in water, about 30 times as sweet as refined cane-sugar, however, sweetness is still easily perceptible at a dilution of 1:10,000 in comparison to sugar 1 : 140; saccharin 1 : 50,000. pH of 10% aqueous solution is 6.5 ± 1.0. It is practically insoluble in alcohol, ether, benzene, $CHCl_3$.

The Calcium salt, also called Cyclan, Sucaryl calcium is shown in Fig. 3.48; its mol. wt. is 396.54. It is dihydrate, crystals with pleasant, very sweet taste, freely soluble in water, practically insoluble in alcohol, benzene, chloroform, ether. pH of 10% aqueous solution is same as of sodium salt above, i.e. 6.5 ± 1.0. It is understood to be more resistant to cooking temperatures than saccharin. Calcium cyclamate is somewhat less sweet than sodium cyclamate.

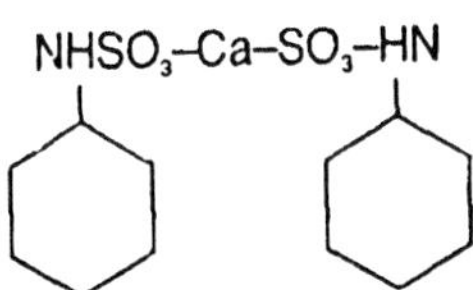

Figure 3.48

Cyclamate and saccharin in the ratio of 10:1 have synergistic, sweetening properties and improved taste.

FDA cancer assessment committee ultimately found cyclamate devoid of any carcinogenicity at any organ/tissue site to either sex of any animal species tested. Consequently JECFA and SCF have independently allocated an ADI of 11mg/kg body weight/day for cyclamate.

Cyclamate is produced by the sulfonation of cyclohexylamine. Today the major producers of cyclamate are located in Brazil, Taiwan, China, Indonesia and Spain.

Detection 2 g of $BaCl_2$ is added to 100 ml of sample as prepared aqueous extract, reactants are kept aside for 5 minutes and then filtered. The filtrate is acidified with 10 ml of HCl, 200 mg $NaNO_2$, is added and contents are warmed on a hot plate. A white precipitate of $BaSO_4$ indicates positive test.

Estimation Cyclamate is hydrolysed by acid under pressure to cyclohexamine, which is extracted with $CHCl_3$ and treated with ethanolic p-benzoquinone to form a coloured product which gets absorbed at 493 nm.

Reagents

(i) ***Cyclamate Na/Ca*** 1 mg/ml solution is prepared accurately in water.

(ii) ***p-benzoquinone*** 0.3% solution is prepared in absolute alcohol.

Procedure

A known quantity containing 15-30 mg of cyclamate is accurately weighed from a well homogenised sample into a beaker. In another beaker 10 ml of (i) is pipetted. Both are diluted to 40 ml with water. 13 ml of 6N HCl is added and diluted with water to 60 ml. Each beaker is placed inside a 400 ml beaker. The 400ml beaker is covered with a watch glass and autoclaved for 7 hours at 15 psi (123 ± 2°C). Hydrolysis can be achieved by adding 5 ml each of Conc HCl and 30% H_2O_2 to the sample solution and keeping the flask in boiling-water bath for 2 hours. Contents are transferred into 250 ml separating funnel, pH is adjusted to 12 using 10% NaOH solution and then extracted with 3 x 25 ml $CHCl_3$. Chloroform extracts are combined, washed till free of alkali, dried over anhydrous Na_2SO_4 and transferred quantitatively to a 100 ml volumetric flask. The volume is made up with $CHCl_3$.

An aliquot of sample and (i) are pipetted into a 50 ml volumetric flask, kept in water-bath maintained at 60°C for 2 hours, protected from direct light. The flask is cooled and volume made up with $CHCl_3$. It is homogenised and absorbance is read at 493 nm in a spectrophotometer, whence cyclamate content is calculated.

ACESULFAME-K

Molecular formulae of Acesulfame is $C_4H_5NO_4S$, molecular wt. 163.15, potassium salt $C_4H_4KNO_4S$. It is also called acesulfame-

K, Sunette, mt. pt. 250°C, very soluble in water, DMF, DMSO, soluble in alcohol and glycerine-water.

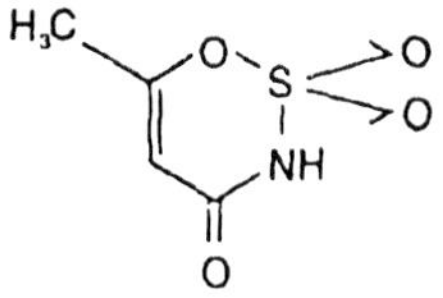

Figure 3.49

Though acesulfame itself is a non-nutritive synthetic sweetener, commonly it is the K-salt which is used as sweetener for foods and cosmetics. It is an approved additive in about 80 countries including India. Because of its characteristics, its ADI (acceptable daily intake) is up to 15 mg/kg of body weight.

Acesulfame-K is excreted completely unmetabolised, hence is non-caloric, does not interfere or change the normal functions of the human body, its average half shelf life is 2.5 hours or even less. It is therefore suitable for diabetics, individuals having sucrose intolerance and also others. Since it is 200 times sweeter than sucrose, therefore its daily intake is equivalent to 15 x 200/1000 = 3 gm/kg of body weight of sucrose which is practically twice the average sucrose consumption in most countries.

This sweetness of 200 times is in diluted aqueous solutions. Sweetness intensity decreases with increasing concentration to values above 100 times. Sweetness intensity factors depend on factors like concentration, temperature, pH level and the composition of the product in which the sweeteners are used.

In most foods and beverages, the taste characteristics of acesulfame-K are distinguished by a fast onset of sweetness without noticeable delay. Its sweetness does not linger and fades away quickly. It has good synergy when blended with aspertame and cyclamate, but not with saccharin. As aspertame's sweetness shows a slightly delayed onset therefore with acesulfame-K, the time intensity profile gets modified. In other words, acesulfame-K renders immediate sweetness and aspertame's delayed onset makes the user feel a quick and lingering sweetness liked by one and all.

Acesulfame-K is a white, crystalline material, stable at room temperature. It decomposes on heating without melting. At 20°C more than 20% (w/w) solution can be prepared and at 100°C it is 50% (w/w) . Since the aqueous solutions are stable hence can be stored also conveniently.

Detection (TLC detection of acesulfame, saccharin and cyclamate.)

Reagents

(i) Polyamide
(ii) 2, 7-dichlorofluorescein
(iii) Bromine
(iv) Formic acid
(v) NH_3 5%
(vi) Xylol
(vii) Propanol
(viii) Methanol
(ix) Ion-exchange resin: Amberlite LA-2.

Procedure

The sweetener is extracted from acidified food product with water or acidified aqueous extract is passed through (ix) and washed with water. Sweeteners are eluted with (v), elute solution is dried under vacuum and the residue dissolved in (viii). Alternatively, these sweeteners are extracted from acidified sample, pH 0.6, with ethyl acetate, using concentrated ethyl acetate for TLC.

6 ± 4 ul of sample solution is applied along with standards on TLC plate coated with (i). The plate is developed to about 15 cm height with a developing solvent consisting of xylol: n-propanol: iv :: 5 : 5 : 1, the plate is dried in an air current and sprayed with 0.2% solution of dichlorofluorescein, again dried and is examined under UV lamp (360 nm). In daylight identification, the plate is placed in a chamber containing bromine then exposed to ammonia vapours. Spots appear on a reddish background.

Determination (HPLC)

Reagents

(i) ***Mobile phase:*** Methanol : H_2O :: 10 : 90, adjusted to 0.1 M using tetrabutyl ammonium sulphate.

(ii) ***Standard acesulfame molution:*** 0.1 mg/ml in water.

Conditions

(a) HPLC with a UV detector at 227 nm.
(b) Lichrosorb-RP 18 (10 μm).
(c) Pressure-160 bar
(d) Flow rate-40 ml/hour
(e) Temperature-ambient
(f) Sample volume 15 ± 5 ul.

Preparation of Sample

A. ***Liquid samples*** Liquid samples like juices are filtered through 0.45 mm filter (mfr: Milipose INC) and (f) Injected.

B. ***Solid samples*** 10 g of the sample is vigorously stirred with 100 ml water for 30 minutes and centrifuged. An aliquot of this solution is passed through 0.45 mm filter as before, first few drops are discarded, while the rest of the filtrate is collected and chromatographed.

Procedure

Standard (ii) above ranging from 5-20 ul is injected and peaks are recorded. Peak area is calculated and a calibration graph is drawn using ug of substance vs peak area. Sample solutions ranging from 10-20 ul are injected and the sample peak area is recorded. Acesulfame content is calculated from its peak area and the aforesaid calibration graph is made.

DULCIN

Dulcin, (4-Ethoxyphenyl) urea; p-phenetylurea $NHCONH_2$ is known in the trade as Sucrol and Valzin. Its mol. formula is $C_9H_{12}N_2O_2$, its mol. wt. is 180.20 and molecular structure is as given in Fig. 3.50.

It is made by treating p-phenetidine with phosgene and then with ammonia or urea.

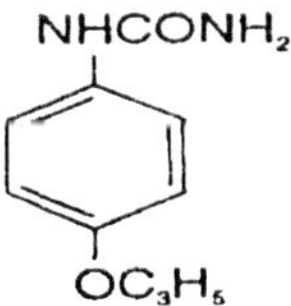

Figure 3.50

Dulcin has lustrous needles, is very sweet in taste-about 250 times as sweet as cane sugar, mt. pt. is 173-174°C. It is soluble in 800 parts of cold water but only 50 parts of boiling water and 25 parts of alcohol. It is a non-nutritive sweetener.

Detection

Preparation of the sample 100 ml of sample is made alkaline with 10% NaOH solution or an alkaline aqeuous extract is prepared. In either case, the solution is extracted with 3 x 50 ml portions of diethyl ether and then divided equally into three porcelain dishes. The solvent is allowed to evaporate at room temperature and the residue is finally dried.

Tests

(i) ***Deniges-Tourrou test*** The dried first residue is moistened with HNO_3 and one drop H_2O. Orange-red coloured precipitate indicates presence of dulcin.

Note: Vanillin, if present, interferes in this test.

(ii) ***Laporola-Mariani test*** The second residue is exposed to HCl gas for 5 minutes, one drop of anisaldehyde is added. Blood red colour means dulcin is present.

(iii) ***Dimethyl amino benzaldehyde test*** To the third residue this reagent's one drop is added. A brick-red colour is deemed a positive test. The reagent is prepared by dissolving its 1 g in 10 ml HCl and made up to 100 ml.

Estimation

UV-spectrophotometric method Dulcin is extracted from the prepared sample under alkaline conditions with diethyl ether. The residue after removal of solvent is dissolved in ethyl acetate and its absorbance is read at 294 nm.

Procedure 50 g of the sample is transferred into a 250 ml separating funnel and made alkaline with 10% NaOH solution. It is extracted with 4 x 100 ml portions of diethyl ether and shaken for 2 minutes each time. The extracts are combined and washed with 10 ml water. The water layer is discarded. Ether is evaporated and the residue is dried at 110°C for 30 minutes then dissolved in 50 ml ethylacetate and transferred quantitatively into a 100 ml volumetric flask, the volume is then made up. If required further dilutions may be done. Absorbance is read spectrophotometrically at 294 nm, ethyl acetate being the blank.

Standard solutions of various dilutions of dulcin are made in ethylacetate. The absorbance/concentration graph is prepared and from it the amount of dulcin is computed.

SUCRALOSE

Sucralose has high quality of sweet taste, stability during processing and storage conditions used for food and beverages, low acute toxicity and high water solubility. Syrup's $[\alpha]_D$ is + 68.2° (*c*= 1:1 in ethanol). Anhydrous crystalline form is orthorhombic, needle like crystals, mt. pt. 130°C. LD_{50} in mice > 16 g/kg. It is roughly 600 times sweeter than sugar with an approximate range of 400-800 times depending upon the level of sweetness desired and the individual product application.

In 1976 the possibility of enhancing the sweetness of sucrose chemically was discovered. Prior to this all derivatives of sucrose that had been tasted had been found to be tasteless or bitter. Hough, *et al* (1976) showed that substitution of the hydroxyl groups in sucrose resulted in a significant increase in the perception of sweetness. This discovery stimulated an extensive research. Consequently hundreds of compounds were prepared in a classic structure-activity relationship study. The factors important for intensification of the sweetness of sucrose were found to be the presence of hydroxyl groups at the 2, 3, 6 and 3^1 positions, with large, highly electronegative groups at the remaining positions. The sweetest compound discovered was the 4, 1^1, 4^1, 6^1–tetrabromo compound, approximately 7500 times sweeter than sugar itself.

A selection of some of the sweet compounds produced and their relative sweetening powers are listed in Table 3.3.

Table 3.3

Derivatives of sucrose	*Sweetening power (X sucrose)*
4-chlorogalactosucrose	5
1^1 -chlorosucrose	20
6^1-chlorosucrose	20
6,1^1, 6^1-trichlorosucrose	25

Derivatives of sucrose	*Sweetening power (X sucrose)*
4, 1^1, 6'-trifluorogalactosucrose	40
1^1, 6^1-dichlorosucrose	80
4, 1^1-dichlorogalactosucrose	120
4, 1^1, 6^1-triiodogalactosucrose	120
4, 1^1, 6^1-trichlorogalactosucrose	600
4 1^1, 6^1-tribromogalactosucrose	800
4 1^1, 4^1, 6^1-tetrachlorogalactosucrose	2200
4, 1^1, 4^1, 6^1-tetrabromogalactosucrose	7500

As can be seen, the range is very large, with very small changes in the configuration and conformation having a profound effect on the sweetness level.

Sucralose is biodegradable. It is a white and free flowing powder, odourless, having negligible effect on pH of solutions, has zero calorie content, and cannot support the growth of streptococcus mutans in the oral cavity. It is also non-carcinogenic.

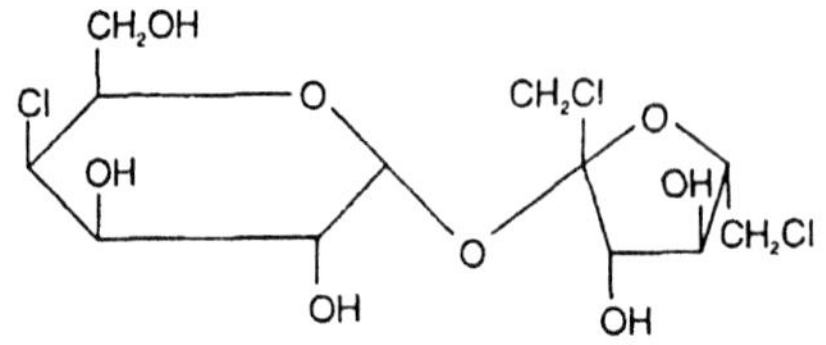

Sucralose

Figure 3.51

Under extreme conditions of acid and heat, sucralose will slowly hydrolyse to its component monosaccharide derivatives. But the chlorine atoms have the effect of stabilising the glycosidic linkage towards protonation thereby reducing the rate of this reaction by approximately two orders of magnitude. Practically the loss of sucralose by acidic hydrolysis in food systems is negligible. Sucralose-sweetened soft drinks show no loss of sweetness after storage for 1 year at 20°C, therefore consistent products can be manufactured and marketed with long shelf lives without degradation and consequential loss of sweetness. Sucralose can withstand high temperature of UHT and pasteurisation processes, and can even be used in cooking, baking and extrusion.

Thus highly stable sucralose is compatible with all common food ingredients, as it does not undergo any interactions with any colours, flavours or other components. Unlike some sweeteners which have a delayed onset of sweet taste or are characterised by a long, lingering sweet taste, sucralose is close to ideal, with a rapid onset of sweetness perception and a persistence similar to that of sugar. Further, the flavour profile of sucralose being similar to sugar, it does not give rise to pronounced bitter, metallic or astringent sensations sometimes associated with other low calorie sweeteners.

Analysis of Sucralose in Beverages

The determination of sucralose in beverages is typically performed by HPLC analysis with refractive index detection. Generally, some form of sample clean up is required prior to analysis to separate and remove potentially interfering components, and this is often achieved via solid phase extraction cartridges. The packing for these cartridges that have proved to be most effective are C_{18} reverse phase silica and Alumina. The C_{18} cartridges adsorb non-polar compounds while polar components pass through unretained. When sucralose is injected onto the cartridge in an aqueous solution, it is retained on the adsorbent while more polar components of the sample matrix are discarded with the eluent. Sucralose can then be eluted from the cartridge by passage of a less polar solvent such as methanol, leaving the non-polar components on the adsorbent. This procedure also allows a concentration step to be incorporated- for example, if a 10 ml aliquot of the sample is injected onto the cartridge and the sucralose is eluted with 3 ml of methanol, a 10/3 times concentration is achieved.

The alumina cartridges are available in three forms; (a) alumina A (acid), alumina N (neutral) and alumina B (basic). In general the Alumina A cartridge has proved to be the most suitable for use with sucralose-containing beverages although some applications for the alumina N form have been found (e.g. black currant beverages). Sucralose passes through these columns in both aqueous and methanolic solutions while colours and basic/acidic components are retained on the packing.

The following guidelines give details of the techniques that can be applied for each of the main product types (i.e. soft drinks, milk based products and alcoholic beverages).

Soft Drinks

1. Soft drinks, whether still or carbonated, can be treated in a similar manner although all carbonated beverages must be de-carbonated prior to analysis by adding approximately 20 ml of sample to a 100 ml glass beaker and placing in an ultrasonic bath for about ten minutes.
2. Simple products, containing only water, flavourings and sweeteners (including sugars) often require no further treatment other than filtration through a 0.45 μm filter, to remove particulate material prior to HPLC analysis[5]. Problems can be encountered, however, with low pH products (below 4) in the form of a broad triangular shaped peak which can interfere with the sucralose peak. In this case, the sample should be passed through an Alumina A cartridge before filtration[2].
3. Where samples are highly coloured or contain gums, starches or other proprietary products to improve the body of the product, cleanup using an alumina A solid extraction cartridge (Waters Associates) is normally required. Approximately 5 ml of sample should be slowly passed through the cartridge, the first 2 ml being discarded. The remaining sample should be collected and filtered through a 0.45 μm filter prior to analysis. The alumina A cartridge is particularly efficient in removal of caramel colour from cola beverages. Possible problems that may be encountered however, include build up of pressure on the HPLC column, interference of sample components with the sucralose peak, or lack of sensitivity. Build up of column pressure indicates that the alumina A cartridge is not removing a sample component (such as CMC) that subsequently fouls the analytical column. In such cases, clean up using C_{18} solid phase extraction cartridge maybe helpful. This approach also allows a concentration step to be incorporated and should therefore be used if sensitivity is a problem[5]. Elimination of the interferences can

be achieved either by modification of the HPLC conditions[5] and/or by using alternative extraction cartridges such as alumina N or C_{18}.[4,5]

CMC means carboxymethyl cellulose.

4. If the passage of the sample through an alumina A cartridge has not removed all interfering components, particularly from a highly coloured sample, the use of an alumina N cartridge should be tried. To clean up the sample using an alumina N cartridge, approximately 5 ml is passed through the column, the first 2 ml of the eluent is discarded. The remaining sample is collected and filtered through a 0.45 μm filter prior to analysis. This approach has been found to be particularly successful in the case of a black currant beverage, where passing the sample through an alumina N column has been found to give a clearer extract then that achieved using an alumina A cartridge.
5. The C_{18} reverse phase cartridge Sep Pak (Waters Associates)/ Bond Elut (Anachem) achieve both sample clean up and concentration of aqueous solutions. The C_{18} cartridge is activated by passage of 3 ml of methanol followed by 2 ml of water; 5-10 ml of sample is slowly injected onto the column and the eluent is discarded. The column is then rinsed by injecting 5 ml water. Finally sucralose is eluted from the cartridge with 3ml methanol. The methanol extract should be taken to dryness under a stream of air at room temperature and the residue redissolved in an appropriate volume of water or HPLC eluent. The volume of sample placed on the column depends on the nature of the sample and the capacity of the cartridge. In the wake of column pressure build up during the HPLC analysis, it is possible that the cartridge has been overloaded, thus reducing the clean-up's efficiency. The way out is that less sample should be added to the C_{18} cartridge by diluting the beverage with water prior to injecting 5-10 ml of the diluted sample onto the cartridge.
6. In some circumstances, where the above procedures fail to remove interfering components, it may be necessary to pass the sample through two extraction cartridges to achieve a satisfactory cleanup. For instance, a sample may be passed

through an alumina A cartridge before being injected onto a C_{18} cartridge as illustrated below (7).

7. Beverages containing CMC have been found problematical because any CMC remaining in the final extract will foul the HPLC column and cause a significant pressure build up so that the sample should be diluted to iesser CMC concentration to about 0.02% or less and then passed through an alumina A column and a portion of this extract further cleaned up by means of a C_{18} cartridge.

Milk Based Products

1. Protein removal is essential prior to the HPLC analysis. It is generally done by pipetting a 5 ml sample aliquot dropwise into 5 ml of methanol which will precipitate the proteins. Slow sample addition is important to avoid getting sucralose precipitated. The resulting solution is allowed to stand for a minute before centrifugation at about 5000 rpm for 20 minutes. The supernatant is then further cleaned up through a solid phase extraction cartridge. The alumina A cartridges are particularly suitable for samples containing fats and lipids which are retained on the cartridge absorbent while sucralose passes through. In this way most milk products can be successfully cleaned up by passing the deproteinated solution through an alumina A cartridge. However, samples containing stabilizers, thickeners, etc. may need to be treated through a C_{18} reverse phase cartridge (Waters Associates) as well as, or instead of, the alumina A cartridge. As the deproteinated supernatant comprises 50% water and 50% methanol, sucralose passes straight through the C_{18} cartridge while less polar components are retained on the absorbent. Either way, the first 2 ml of sample eluting from the cartridge needs discarding. The extracts from the solid phase extraction cartridges should then be filtered through a 0.45 μm filter prior to HPLC analysis.
2. Protein precipitation can be achieved by adding 30 ml ethanol to about 2.5 g of sample, placing the mixture in a 60°C water bath for about 18 minutes and then in an ultrasonic bath, at room temperature, for 10 minutes. The solution is centrifuged and an accurately measured volume of the supernatant is dried. The dried residue is dissolved in the HPLC eluent,

filtered and analysed. If further cleanup is needed, the residue can be redissolved in water and then passed through either an alumina A or C_{18} solid phase extraction cartridge. As this extract is aqueous, sucralose is retained on the C_{18} cartridge permitting a concentration step to be included if required like in soft drinks, above.[5]

3. Extracts containing 50% methanol can be successfully chromatographed under the standard HPLC condition[5]. Higher levels of methanol or solutions having more than 5% ethanol adversely effect the chromatography. Under these conditions the alcohol concentration is reduced to an appropriate level prior to analysis.

Alcoholic Beverages

1. All carbonated beverages are invariably decarbonated prior to analysis, normally by adding about 20 ml of sample to a 100 ml glass beaker and placing it in an ultrasonic bath for ten minutes. In the case of alcoholic beverages, it is important to take an accurately measured volume of sample for decarbonation. The decarbonated sample is backed up to a known volume. This corrects for any loss of ethanol through evaporation.
2. Sample is generally cleaned up by alumina A cartridge, although in some cases of black currant beverages the alumina N cartridge yields better sample clean up. About 5 ml of sample is slowly passed through the cartridge, first 2 ml is discarded. The remaining sample is collected and filtered through a 0.45 μm filter prior to HPLC analysis.
3. ***HPLC analysis*** Analysis of sucralose in extracts from food products is usually done by HPLC with refractive index detection. Typical conditions are given in Table 3.4

Table 3.4

1.	Column	5 μm C_{18} Resolved Red Pak of Waters Associates
2.	Eluent	H,O : methanol :: 7 : 3
3.	Flow rate	1.25 ± 0.25 ml/minute
4.	Detector	Refractive index (RI)

Regarding columns, other reverse phase columns are also suitable although the aforesaid has the advantage of accommodating injection volumes up to 150 ml, if needed.

The exact eluent composition depends upon the nature of the sample and the column type being used. The above composition, however, gives a good "starting point" and can be adopted as necessary to provide the required solution. In general, a methanol content of 30 ± 10% has been found satisfactory.

Flow rate is also dependent upon the nature of the sample, the column and the eluent composition.

Sucralose can be determined by ultraviolet (UV) absorption at 190 nm or refractive index detection and gives a linear response over the range 50 ppm to 10,000 ppm by both techniques. Refractive index is, however, the preferred method for the analysis of sucralose in most food matrices because of high background interferences experienced with UV at 190 nm. Mass detectors and evaporative light scattering detectors can also be used for the determination of sucralose and offer the advantage of allowing gradient elution. However, these detectors are less readily available than the RI detector and problems can be encountered with the high proportion of water in the eluent used for sucralose analysis.

The detection limit is dependent upon the make and model of detector used and on the type of integration system available. The Water's model 410 RI detector can be used to quantify sucralose accurately, in solution, down to concentrations of 50 ppm with a standard integrator and lower concentrations (*C* 1-10 ppm) with a computerised integration system.

Method validation Determination of recovery factor for the extraction and the reproducibility of the method must be done before using a method as a routine method by spiking a blank (i.e. unsweetened) formulation of the product within a known amount of sucralose. 8 ± 2 portions are taken through the extraction procedure, the level of sucralose quantified against external standards. The recovery factor can be determined from these values and the spread of results gives an indication of the reproducibility of the method, i.e. co-efficient of variation. It is essential to analyse a blank sample to check that there are no interfering components co-eluting with the sucrolose peak. Some examples of recovery factors are:

(i) ***Carbonated cola*** formulated with xanthan gum and containing 159 ppm sucralose. Samples were cleaned by passing through Alumina A cartridge prior to analysis. Recovery 99 + 0.5%

(ii) ***Carbonated lemon/lime*** formulated with CMC and containing 118 ppm sucralose. Samples were diluted and then treated through an Alumina A cartridge followed by a C_{18} cartridge to achieve a satisfactory clean up. A recovery factor of 97% + 3.1% was determined.

(iii) ***Strawberry milk*** containing 60 ppm sucralose; following deproteination, the samples were passed through alumina A Sep Pak cartridges. A recovery of 100% + 3.6% can be achieved.

Analysis of Sucralose in Cereal Products

Prior to extraction, the representative sample is ground to a fine particle size. Extraction of sucralose is achieved by adding hot (near boiling) water to a weighed portion of the ground sample and shaking the mixture vigorously for 10-15 minutes. Once cool, the sample is made up to a known volume using water at room temperature. The mixture is centrifuged at about 5,000 rpm/filtered through glass fibre filter paper (Whatman) to remove the undissolved material. The supernatant/filtrate is further cleaned up prior to HPLC analysis, by means of C_{18} solid phase extraction cartridge.

Solid phase extraction Aqueous extracts of all baked and extruded products are cleaned up using a C_{18} Sep Pak cartridge. A 5-10 ml aliquot of the supernatant/filtrate is injected on the above cartridge, previously activated by passage of 3 ml of methanol followed by 2 ml of water. The cartridge is washed with 5 ml of water and finally sucralose is eluted with 3 ml of methanol. The methanol extract is dried under a stream of air at room temperature then redissolved in 3.0 ml of water. The solution is filtered through a 0.45 um filter, prior to HPLC analysis. This procedure recovers 100% sucralose.

HPLC analysis, analytical column, eluent, flow rate, detection and method revalidation are precisely same as already detailed under alcoholic beverages para (3).

Analysis of Sucralose in Canned Products

1. ***Sample preparation*** Canned products are normally sweetened by adding sucralose to the syrup or sauce, prior to canning. On standing, the juice is absorbed by the product dispensing sucralose through out the contents of the can. Homogenisation of the test material is important for ensuring an accurate sucralose determination. This is best achieved by taking the entire contents of a can and macerating them with an eleotric blender.
2. ***Aqueous extraction*** Sucralose extraction is achieved by shaking a weighed portion of the homogenised sample with water for 5-10 minutes. The mixture is centrifuged at about 7,000 rpm to separate pulp from the solution. The supernatant is further cleaned up prior to HPLC analysis with the help of a solid phase extraction cartridge.
3. ***Solid phase extraction*** The C_{18} reverse phase packing Sep Pak cartridges (Waters) or Bond Elutes (Anachem) (*) can be used to remove components that are either more polar or less polar than sucralose and in addition allows, a concentration step to be incorporated. The Alumina columns (*) remove colours, acidic as well as basic components. A few successful examples are:

(a) An extract from canned apples can be passed through Alumina A cartridge to achieve satisfactory clean-up. Nearly 5 ml of the supernatant is slowly injected into the Alumina A Sep Pak. The first 2 ml of the eluent is discarded, the rest is collected and filtered through a 0.45 μm filter, prior to HPLC analysis.

Samples containing CMC/polydextrose must be treated through a C_{18} Sep Pak cartridge to remove these components prior to HPLC analysis. A 5 ml aliquot of the supernatant is injected onto C_{18} cartridge, preactivated by passage of 3 ml of methanol followed by 2 ml of H_2O. The cartridge is washed with 5 ml water to remove polar components and sucralose is then eluted with 3 ml methanol. The methanol extract is dried under a stream of air at room temperature and the residue is redissolved in 3.0 ml water. Finally the solution is filtered through a 0.45 μm filter to remove any particles, prior to HPLC analysis. Presence of CMC/polydextrose in the extracts results in a build up of back pressure on the HPLC column.

An alternative extraction method for jams containing CMC/ polydextrose is to extract them with methanol. Sucralose dissolves in methanol, leaving CMC, etc. with the bulk of the sample. After centrifugation, about 10 ml of the supernatant is passed through a Alumina A Sep Pak cartridge, first 2 ml of eluent is discarded, the rest is completely eluted, a 5.0 ml of aliquot is accurately pipetted into a clear vial. The methanol extract is dried, the remaining procedure is same as described earlier.

(i) Ready to eat desserts, dairy desserts like custard-whipped topping/ready to eat pudding are extracted with methanol/ethanol, extract is passed through an Alumina A Sep Pak cartridge. The sample is thoroughly mixed, 2 g are accurately weighed into 10 ml volumetric flask which is partially filled with methanol, contents shaken vigorously for about 2 minutes. Samples difficult to disperse in cold are either warmed or the flask is placed in an ultrasonic bath to achieve satisfactory mixing. Only after shaking the sample volume is made up with methanol. The contents are transferred to a suitable tube, centrifuged at about 5000 rpm for 5 minutes. Nearly 10 ml of the supernatant is passed through an Alumina A Sep Pak cartridge, first 2 ml of eluent is discarded, the rest eluent is collected and a 5.0 ml aliquot is accurately pipetted into a clean vial. This extract is dried under an air stream at room temperature, the residue redissolved in 3.0 ml of water, prior to HPLC analysis.

(ii) Dry mix products like gelatin dessert and starch based puddings belong to this class. Their dry sample is extracted with methanol, the mixture is centrifuged and the supernatant is passed through a solid phase extraction cartridge. Accurately weighed sample of about 1 gm is mixed with 10.0 ml of methanol and elution achieved by passing an aliquot of the supernatant (about 5 ml) slowly through an Alumina B cartridge. The eluent is filtered through a 0.45 μm filter and analysed by HPLC.

(b) In the case of canned peaches, satisfactory clean up has been attained using a C_{18} Sep Pak cartridge. A 5 ml aliquot of the supernatant is injected on to C_{18} Sep Pak that has been previously activated by passage of 3 ml of methanol followed by 2 ml of water. The cartridge is washed with 5 ml of water and sucralose is finally eluted with 3 ml of methanol. The methanol extract is dried under a stream of air (at room temperature), the residue redissolved in 3.0 ml of water. The solution is filtered through a 0.45 μm filter prior to HPLC analysis.

(c) To obtain adequate clean up of extracts from baked beans (beans in tomato sauce) it is necessary to use two solid phase extraction cartridges. About 5 ml of the supernatant is passed through a 5 mm filter onto an Alumina N cartridge. The eluent is collected and an accurate measured volume (5ml) is injected onto an activated C_{18} cartridge. The C_{18} Sep Pak is washed with 4 ± 1 ml of water and sucrolose is eluted with 3 ml of methanol. The methanol extract is dried under a stream of air, the residue is dissolved in 3.0 ml of water. The resulting solution is filtered through a 0.45 μm filter and analysed by HPLC.

Recovery of sucralose from the sample matrix in the above three cases (a to c) is nearly 100%.

Like in (c) earlier, the HPLC analysis, analytical column, eluent, flow rate, detection and methanol revalidation are precisely same as already detailed under alcoholic beverages para (3).

Analysis of Sucralose in Yoghurts, Jams and Desserts

(i) **Yoghurts** Yoghurt samples are generally homogenised in a blender or liquidizer prior to analysis. A weighed portion of the homogenised sample is then extracted by dispersing it in water (say, 10 g to 100 ml) and mixed thoroughly. To attain sample clean up, a 5-10 ml of aliquot of the aqueous extract is passed slowly through an Alumina A Sep Pak cartridge. First 2 ml of this eluent is discarded, the rest collected and filtered through a 0.45 μm filter, prior to HPLC analysis. If the yoghurt sample contains too much of fruit pulp, it must be centrifuged or the aqueous extract, coarse filtered before it is injected onto the Sep Pak, to avoid blocking of the cartridge.

(ii) **Jams** Most jam samples can be extracted in water and then cleaned up by means of an Alumina A or C_{18} Sep Pak cartridge. 2.5 g of the sample is weighed into 25 ml volumetric flask, about 10 ml of water is added, contents shaken vigorously for 10 minutes. For sample dispersion the solution is either

warmed or hot water is used if required to achieve satisfactory extraction. The sample is shaken, the volume made up with water, contents mixed and centrifuged at about 5000 rpm for nearly 10 minutes. The supernatant must be further cleaned up via a solid phase extraction cartridge.

Aqueous extracts from jam samples, free of CMC/polydextrose methanol, are shaken vigorously for about a minute. The mixture is centrifuged at 2000 rpm for about 5 minutes.

There is another group like orange gelatin dessert where the supernatant does not need further clean up. Therefore, an accurately measured volume can be dried under a stream of air, the residue treated as stated earlier for final HPLC analysis. But extracts from products like chocolate starch dessert, need to be passed through C_{18} Sep Pak cartridge. As the sucralose is in a methanolic condition, it will not be retained on the column and will pass through the cartridge while less polar components get retained on the absorbent. The alumina columns may also be suitable for clean up of some samples. Sucralose will be eluted unretained while colours and basic/acidic components are held on the cartridge packing. Irrespective of whether C_{18} or Alumina cartridges are used to achieve sample clean up, approximately 10 ml of the methanol extract must be passed through the solid phase extraction cartridge, the first 2 ml being discarded. The remaining eluent is collected and a 5.0 ml aliquot is accurately pipetted into a clean vial, it is dried under an air stream and proceeded as detailed earlier.

The HPLC analysis, analytical column, eluent, flow rate, detection and methanol revalidation are precisely same as already detailed under alcoholic beverages para (3).

References

1. E Oertly and RG Meyer, *J. Amer. Chem. Soc.*, 1919, 2, 145-153.
2. RS Shallenberger and T.E. Acree, *Nature*, 1967,216,480-482.
3. LB Keir, *J. Pharm. Sci*, 1972, 61,1394-1397.
4. RS Shallenberger and M G Lindley, *Food Chem.*, 1977,2, 143-153.

5. A Vander Heijden, LBP Brussels and HG Peter, *Food Chem.*, 1978,3,207-211.
6. JM Tinti and C Nofre, *Sweeteners, Discovery, Molecular Design and Chemoreception,* 1991, pp 88-99.
7. EM Acton, MA Leaffer, S.M. Oliver and H. Stone, *J. Agri. Food. Chem.* 1970, pp 18, 1061-1068.
8. TH Grenby (Ed) Advances in Sweeteners chapter *"Molecular Structure and Sweet Taste"* by WJ Spillane, Blackie Academic and Pr ,fessional
9. *Fourth Annual Report on Carcinogens,* NTP 85-002, 1985, p. 179.

Chapter 4

Food Colours

INTRODUCTION

Colours play a strong role in food identification and in assessment of their fitness for consumption, therefore foods should be prepared only in their natural colours. Food preparation has two parts. The first part is based on nutritional value of food in its natural colours. In the second part, the food apart from being nutritious should also be tasty and attractive. Thus colour in food and drink products is important for the manufacturer primarily because it is important to his customers. This characteristic has no ethnic/geographic barriers. Fruits and vegetables throughout the world abound in colour. Varieties are chosen for their colour. Cooks roast, toast and bake dishes to give an attractive colour. Wines and spirits are processed and matured to give characteristic colours. Brightly coloured foods are used in recipes for their contribution to appearance. Chefs have to arrange the food keeping in mind their visual appeal.

The colours of man made foods from natural ingredients like tea, coffee, bread, cheese, butter, ham, confectionery have expected flavours or other characteristics which one expects after prolonged use. In this way, colour has become a characteristic associated with many manufactured foods and has gained universal importance.

Colouring matter in foods can be broadly classified into two groups natural and synthetic colours. Natural colours consists of chlorophyll, carotenes, anthocyanins, flavones, annato, cochineal, saffron, turmeric, cardamom, betanin, safflower, kokum, grape skin colours, caramel, etc.

Synthetic colours form a major group of food colours which are classified into synthetic colour dyes and mineral pigments. Synthetic colours/dyes are of importance as they are widely used in different foods. These are further divided as acidic and basic dyes. To regulate and enforce safety about the use of synthetic colours, dyes and pigments, there is a separate part VI comprising of Rules 23 to 31 of the P.F.A. Rules 1955. Under Rule 5, there is the appendix B which lays down physio-chemical specifications of foods and under it at para A.26 and its sub-paras, there are specifications about these food additives in detail.

However, certain unpermitted colours such as Metanil yellow, Rhodamine B, Orange G, Blue VRS, Auramine and certain unidentified water and oil soluble colours often appear in foods in contravention to the rules leading to litigation.

Cochineal The dried female insect, *Coccus cacti (L)* encloses the young larvae. It is found in Mexico and Central America but cultivated in West Indies, Canary Islands, Algiers and Southern Spain. About 155 insects/g contain about 10% carminic acid, about 2% coccerin (a wax) and about 10% fat. The colouring matter is alkali carminate contained only in the fatty parts of the insect and in the yolk of the eggs, to the extent of 12 ± 2%.

It is used to colour food products and toilet preparations. Carmine and carminic acid from the insects are used for the manufacture of red and pink inks and lakes also.

Caramel Burnt sugar colouring, made by heating sugar or glucose, adding small quantities of alkali, alkaline carbonate or a trace of mineral acid during the heating. It is a dark brown, thick liquid, pleasant, bitter taste, odour of burnt sugar, density about 1.35, soluble in water, dil. alcohol, insoluble in benzene, chloroform, ether, acetone, petroleum ether, oil turpentine.

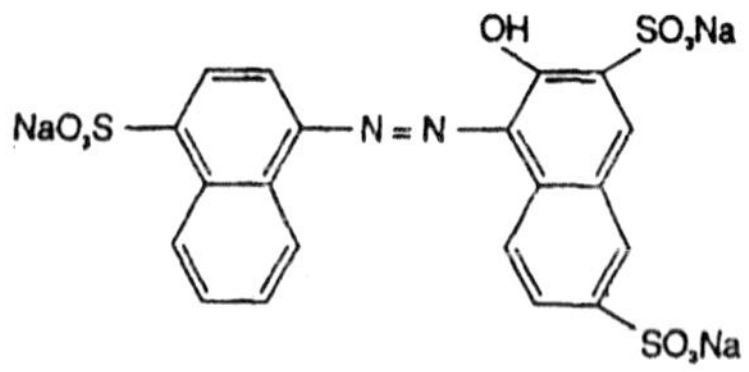

Figure 4.1.

Amaranth

Amaranth is a dark, reddish powder. Absorption$_{max}$ (H_2O) : 522.5 nm 1 g dissolves in about 15 ml of water, also reported as 7.20 g/100 ml of water at 26°C. It is very slightly soluble in alcohol. The aqueous solution is vivid red (1cm layer). HCl does not change the colour intensity of the solution, NaOH increases it. The aqueous solution is stable towards light.

Erythrosine

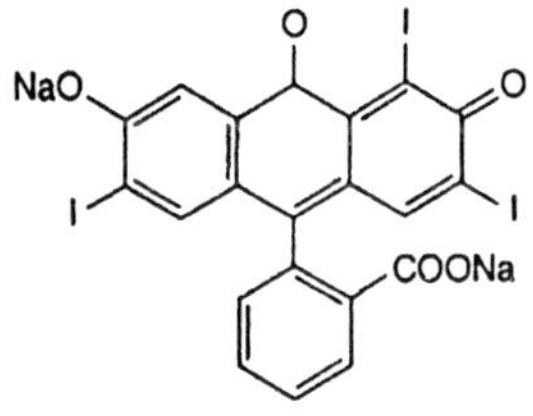

Erythrosine

Figure 4.2.

Brown powder. Absorption$_{max}$ (H_2O): 524 nm; in 95% alcohol 531 nm. Soluble in water to cherry red solution and soluble in alcohol. HCl added to an aqueous solution produces a yellowish-brown precipitate. NaOH produces a red precipitate soluble in excess of the reagent.

Indigo carmine

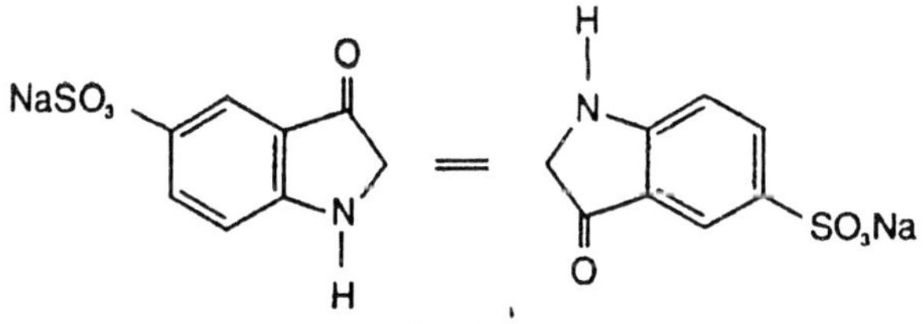

Indigo Carmine

Figure 4.3

Soluble indigo blue, indigotine and acid Blue 74. It is a dark blue powder with coppery lustre and is sensitive to light. Its solutions have a blue or bluish purple colour. 1 gm dissolves in about 100 ml of water at 25°C. It is slightly soluble in alcohol and practically insoluble in most organic solvents. It is also marketed as a paste with water, the dye content varies according to the specification or requirements of the user. It almost always contains $NaCl/Na_2SO_4$ for salting it out. Indigo carmine is very sensitive to oxidising agents. The colour is readily discharged by HNO_3, ClO_3^-. The colour of the aqueous solution fades on standing.

Sunset yellow (FCF)

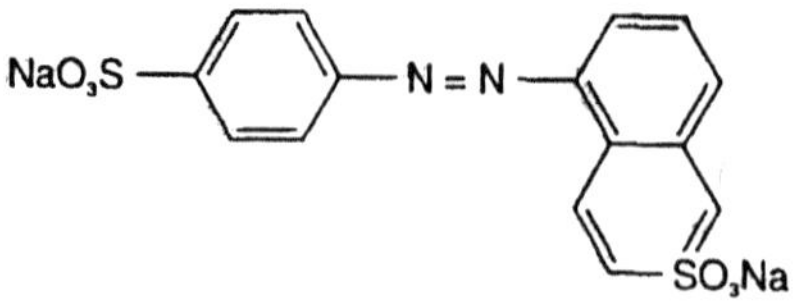

Figure 4.4

It has orange-red crystals. Absorption$_{max}$ (0.02N CH_3-$COONH_4$) : 480 nm. It is soluble in water and is slightly soluble in ethanol. It forms reddish-orange solution in cone H_2SO_4, changing to yellow on dilution.

Tartrazine

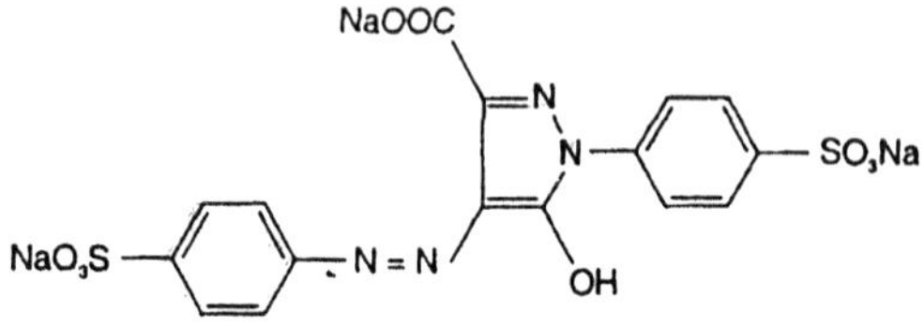

Tartrazine

Figure 4.5

It is bright orange-yellow powder, freely soluble in water. The aqueous solution is not changed by HCl but becomes redder with NaOH.

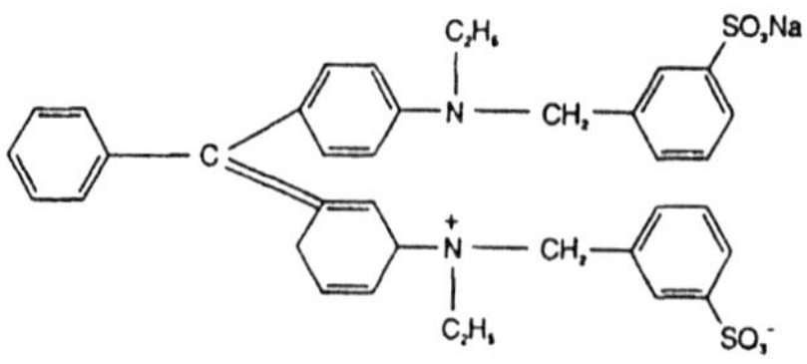

Fast Green FCF

Figure 4.6

Fast green FCF

It exists as dark green powder or granules with a metallic lustre. Its Absorption$_{max}$ is 628 nm. It is very soluble in water and ethanol.

It has dull orange solution in cone. H_2SO_4, changing to dull green on dilution. It has orange solution in conc. HCl or conc HNO_3. Bright blue solution in 10% aqueous NaOH. Dangerous for eyes hence not applicable in eyes area

Brilliant blue FCF

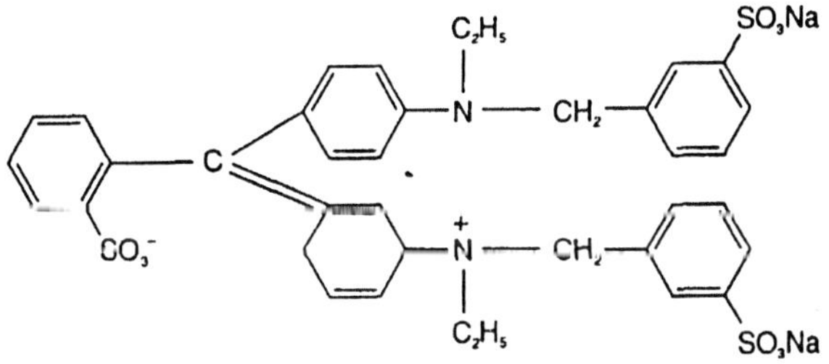

Brilliant Blue FCF

Figure 4.7

It exists as reddish violet powder or granules with a metallic lustre. Its Absorption$_{max}$ is 630 nm. It is soluble in water and ethanol and is insoluble in vegetable oils. It has pale amber solution in conc. H_2SO_4, changing to yellow then greenish blue on dilution. It is dangerous to the eyes hence not applicable in the eyes area.

Carminic acid

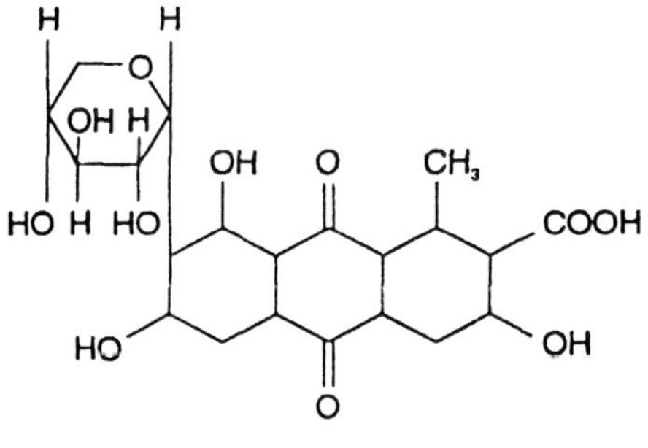

Carminic acid

Figure 4.8

It is the glucosidal colouring matter from the scale insect *Coccus cacti (L),* i.e. cochineal described earlier as well as an essential constituent of carmine. Though it is approved by the PDA for use in foods and drugs but is not permitted under the P.F.A. Rules 1955.

Turmeric (curcumin)

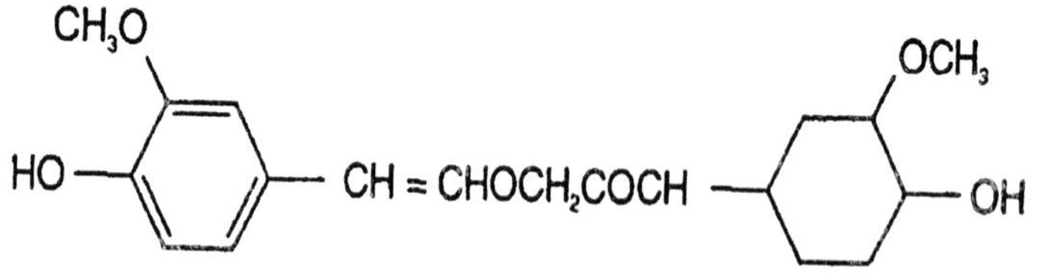

$$\text{HO}-C_6H_3(\text{OCH}_3)-\text{CH}=\text{CHOCH}_2\text{COCH}-C_6H_3(\text{OCH}_3)-\text{OH}$$

Turmeric (Curcumin)

Figure 4.9

Curcumin is an orange yellow crystalline powder soluble in methanol, ethanol, acetone, dichloroethylene and glacial acetic acid and is sparingly soluble in water and ethyl ether. Curcumin in alcohol has yellow colour and light green fluorescence.

It is a colouring matter from the roots of *Curcum longa* (L). Other *Curcuma* species like *C. aromatica, C. zadoria and C. xanthorriza* have very low curcumin content. The mt. pt. of curcumin powder is 183°C. It gives a brownish colour with alkali and a light-yellow colour with acids.

β-carotene:- Its other names are β, β-carotene, carotaben; provatene; solatene. $C_{40}H_{56}$; mol. Wt 536.85. Most important of the pro vitamins A. widely distributed in the plant and animal kingdom. In plants it occurs almost always together with chlorophyll. Isolated first from carrots hence this name.

β-carotene appears as deep-purple, hexagonal prisms from benzene+methanol; red, rhombic almost square leaflets from pet ether, mt. pt. 183°C (evacuated tube). Absorption$_{max}$. ($CHCl_3$); 497, 466 nm. Less soluble than α-carotene. Soluble in CS_2, benzene, chloroform. Moderately soluble in ether, pet ether, oils. 100 ml hexane dissolves 109 mg at 0°C. Very sparingly soluble in methanol and ethanol. Practically insoluble in water, acid, alkalies. Dilute solutions are yellow. Absorbs oxygen form the air giving rise to inactive, colourless oxidation products. Commercial crystalline β-carotene has a vit A activity of 1.67 million USP units/gram. The I.U. of 0.6 μg β-carotene is almost exactly equivalent to 0.3; vit. A. It must be kept tightly closed and protected from lights and stored at low temperature (–20°C).

Canthaxanthin:– β, β-carotene-4-4′-dione; 4,4′-di-oxo-β-carotene; Food Organe 8; carophyll Red ; Orobronze; Roxanthin Red 10; Carotoben plus—all are its other names. $C_{40}H_{52}O_2$; mol. wt 56482. All trans-carotenoid pigment widely distributed in nature. Isolated from the edible mushroom *Cantharellus cinnabarius* (Adams). Also isolated from flamings's feathers.

Violet crystals from methylene chloride, decompose beyond 217°C Absorption$_{max}$ (cyclohexane): 470 ($E^{1\%}_{1\,cm}$ 2250). Soluble in $CHCl_3$, oils. Oil solutions are more red than those of β-carotene.

It is an oral sun tanning agent.

Chlorophyll: It is the well known green pigment of plants. Higher plants and green algae contain chlorophyll a and chlorophyll b in the approx ratio of 3:1. Chlorophyll c is found together with chlorophyll a in many types of marine algae. Red algae (*Rodophyta*) contains principally chlorophyll a and also chlorophyll d.

Riboflavine:–Also known as Vitamin B_2; lactoflavine; Vitamin G; 7, 8-di-methyl-10-(D-ribo–2, 3, 4, 5-tetra hydroxyl pentyl) isoalloxazine; 7, 8-dimethyl-10-ribityliso alloxazine; Beflavine; Flavaxin; Ribica. Its mol wt. 376.36, $C_{17}H_{20}N_4O_6$.

This nutritional factor is found in milk, eggs, malted barley, liver, kidney, heart, leafy vegetables etc. Richest natural source is yeast. Minute amounts present in all plants and animal cells. Occurs in the free form only in the retina of the eye, in whey and in urine; principal forms occurring in tissues and cells are flavine mononucleotide (FMN, riboflavine-5 phosphate) and flavin-adenine di-nucleotide (FAD).

Riboflavine forms fine orange-yellow needles from 2 N acetic acid, alcohol, water or pyridine. It decomposes between 280±2°C but darkens at 240°C. Three different crystal forms having different solubilities in water. When dry, it is not appreciably affected by diffused light, but in alkaline solutions it deteriorates quite rapidly, the deterioration being accelerated by light [α^{25}_{d}] –112° to –122° (50 mg in 2 ml) 0.1 N alcoholic NaOH (diluted to 10 ml with water).

Absorption$_{max}$: 220-225, 266, 371, 444, 475 mm. 1 gram dissolves in from 3,000 to 15,000 ml of water, the variation in the solubility being due to difference in the crystal structure. Slightly more soluble in HCl solutions; less soluble in alcohol than in water (0.0045 g)/ 100ml) of absolute ethanol at 27.5°C. Also slightly soluble in cyclohexanol amylacetate, benzyl alcohol, phenol; insoluble in ether, $CHCl_3$, acetone, bezene; very soluble in dil alkalies (with decomposition).

Although sensitive to alkalies, riboflavine is stable to mineral acids in the dark. Visible or ultraviolet irradiation of alkaline solutions causes the formation lumiflavine q.v. whereas irradiation of acids or neutral solutions gives rise to the production of the blue, fluorescent substance lumichrome q.v. together with varying amounts of lumiflavine.

Annatto:– Its other names known are arnotta and annotta. This colouring matter is present in the seeds of *Bixa orellana* (L); family *Bixaceae*. It contains bixin and several yellow to orange-red pigments which give carotene reactions. It is soluble in alcohol, ether and oils.

Saffron:– Crocus; Spanish saffron; French saffron, its habitat is Western Asia, Kashmir in India, Southern Europe.

Its constituents are volatite oil 1%, picro crocin- a bitter glucoside, crocin (poly chroit), fixed oil, wax.

Saffron is comprised of flattish-tubular, almost thread-like stigmas, nearly 3 cm long, orange-brown colour, strong, peculiar aromatic odour, bitterish aromatic taste.

It is almost entirely used for colouring and flavouring.

For more details refer authors " Manual of Indian Spices" Pub. by Academic Foundation New Delhi-110002.

Metanil yellow

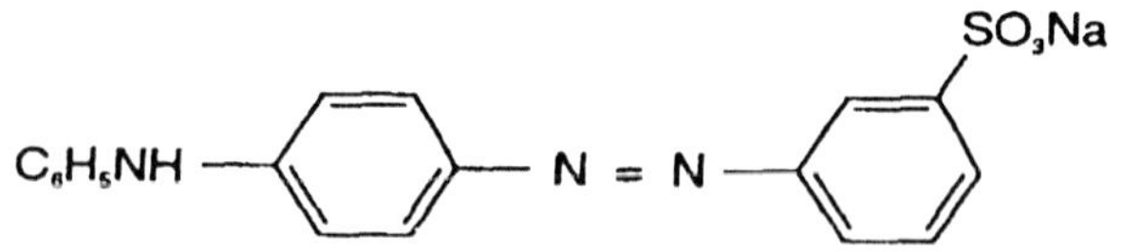

SO_3Na

C_6H_5NH — N = N —

Metanil Yellow

Figure 4.10

It is a brownish yellow powder, soluble in water and alcohol. It is moderately soluble in benzene, ether and slightly soluble in acetone. It is an indicator in 0.1% solution, 2 drops are required for 10 ml liquid. pH:1.2 red to 2.3 yellow.

Orange B

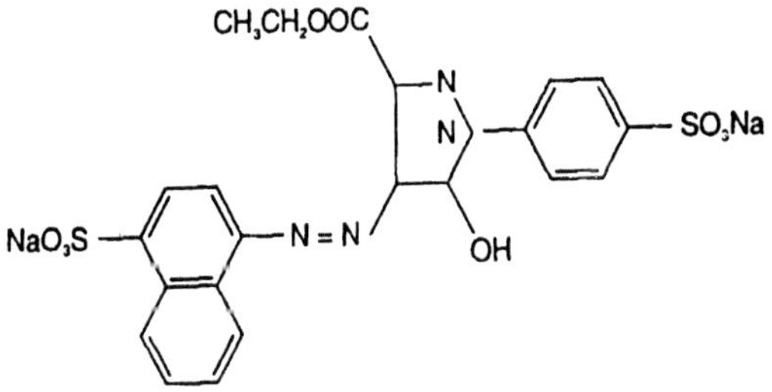

Figure 4.11

It has dull orange crystals. Its Absorption$_{max}$ (0.04N ammonium acetate) : 442 nm. Solubility in water at 77°C is 220 g/litre. Violet solution in conc. H_2SO_4, changing to fuchsia then red on dilution. Red solution in conc. HCl. Yellowish solution in 10% NaOH, changing to brownish yellow on dilution.

Rhodamine B/Tetramethylenediamine

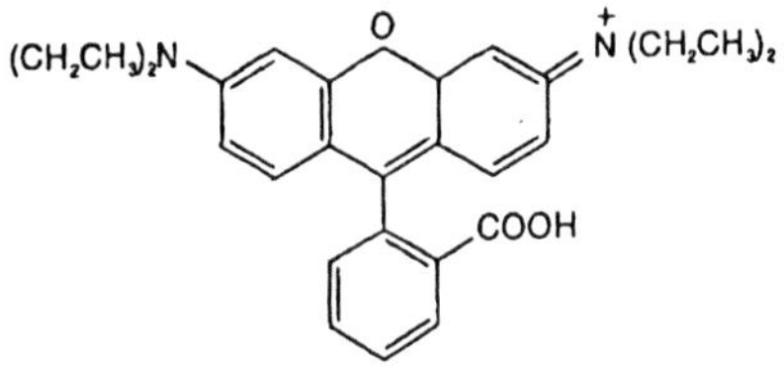

Figure 4.12

It has green crystals or reddish violet powder and is very soluble in water with a bluish-red colour. Its dilute solution is fluorescent. It is very soluble in alcohol and slightly soluble in HCl and NaOH. It is more a reagent for Sb, Bi, Co, niobium, Au, Mn, Hg, Mo, tantalum, Th, W, etc. though provisionally it is listed for use in drugs and cosmetics.

IDENTIFICATION OF NATURAL COLOURS

(*a*) ***Caramel*** It is detected by Fiehe's reaction. The sample solution is extracted with 50 ml ether and is evaporated in a porcelain dish. To the residue, 3 drops of 1% solution of resorcinol in HCl are added. Development of rose colour means caramel is present.

(b) ***Cochineal*** Amyl alcohol solution of the test material is shaken with dil. liq. ammonia. Appearance of purple colour indicates presence of cochineal. Carminic acid present in cochineal gets converted to ammonium carminate which is coloured.

(c) ***Turmeric (curcumin)*** Alcoholic extract of the test material is evaporated almost to dryness on the water-bath with a piece of filter paper. The dried paper is moistured with a few drops of a weak solution of boric acid to which some drops of HCl have been added. The paper is redried. Presence of turmeric/curcumin is indicated if the paper turns cherry red in colour which changes to bluish green by a drop of alkali (NaOH/NH_4OH).

(d) ***Annatto*** Melted fat/oil sample is shaken with 2% NaOH solution, the aqueous extract is poured on a moistened filter paper. The filter paper showing a straw colour is washed with water and a drop of 40% $SnCl_2$ solution is added after which the paper is dried again. If the colour turns purple, the presence of annato is confirmed.

(e) ***Chlorophyll*** The sample is extracted with ether and the ether extract is treated with 10% KOH in methanol. If the extract becomes brown, quickly returning to green, it confirms the presence of chlorophyll.

(f) ***Betanin*** The aqueous suspension is extracted with amyl alcohol. If present, betanin remains in aqueous phase. A cotton piece is mordanted with tannin then dyed in the aqueous phase, appearance of a terra-cotta shade means the presence of betanin.

ISOLATION OF SYNTHETIC FOOD COLOURS

This isolation is carried out with white knitting wool boiled in very dilute solution of NaOH and then in water to free it from alkali. The various solvents required for this purpose are,

(i) NH_3 (Sp.gr. 0.88): H_2O:: 1:99

(ii) NaCl (2.5% aqueous solution and (ii) a 2% in 50% ethanol.

(iii) Dil. acetic acid in water (1:3).

(iv) Isobutanol: ethanol: H_2O :: 1:2:1 (v/v).

(v) n-butanol: H_2O : Glacial acetic acid :: 20:12:5 (v/v).

(vi) Isobutanol: ethanol: H_2O :: 3:2:2 (v/v). To 99 ml of this is added 1 ml of 0.88 sp. gr. NH_3.

(vii) Phenol:H_2O :: 4:1(w/w).

Procedure

Assuming that an acidic colour is present, the interfering substances are removed and the dye is obtained in acid solution prior to boiling with wool.

(a) Non-alcoholic beverages, e.g. soft drinks are usually acidic hence can be directly treated with wool, otherwise, slightly acidified with acetic acid.

(b) Alcoholic drinks are boiled to remove alcohol and acidified if necessary like in (a).

(c) In the case of starch based foods like cakes, custard powder, etc. and candied fruits 10 g of sample is ground thoroughly with 50 ml of 2% NH_3 in 70% ethanol, kept aside for one hour then centrifuged. The separated liquid is poured into a dish, evaporated on a water bath, the residue taken up in 30 ml of dil. acetic acid.

(d) Products with high fat content like sausages, meat, fish paste, etc. are defatted with light petroleum, colour extracted with hot water, acidifying is carried out keeping in mind that oil soluble colours tend to give coloured solutions in organic solvents. In the event of difficult extraction, the sample is treated with 70 ± 20% acetone (warm) or alcohol having 2% NH_3. This shall precipitate the starch. The organic solvent is removed before acidification like (c).

Colour extraction 20 cm length of white woollen thread is introduced in a beaker containing about 35 ml of the prepared acidified solution. It is boiled for a few minutes till the woollen thread is dyed. The thread is taken out and washed with tap water, transferred to a small beaker containing dilute NH_3 and reheated. If the colour gets stripped, presence of an acidic synthetic colour is indicated. After wool removal, the liquid is slightly acidified and a fresh woollen thread is introduced, boiling the contents till this thread gets dyed. The dye is again extracted from this woollen thread with NH_3 like before then filtered through a small cotton plug, the filtrate is concentrated over a hot water bath.

This double stripping technique usually gives a pure colour extract with exceptions. Natural colours may also dye the wood during the first treatment, but the colour is not usually removed by ammonia.

Basic dyes can be extracted by making the food alkaline with NH_3, boiling with wool and then stripping with dilute acetic acid. At present all the permitted water-soluble synthetic colours/dyes are acidic hence an indication of the presence of a basic synthetic colour suggests that an unpermitted colour is present.

Identification by paper chromatography A pencil line, parallel to the bottom edge of the paper (Whatman No.l) is drawn at about 2 cms distance. The concentrated solution of the unknown dye is spotted on the line together with a series of spots (about 2 cms apart) of aqueous solutions of standard permitted dyes of similar colour and dried. The ascending technique chromatogram is run using a selected solvent. Solvent no. v is often helpful for general purposes, but it is better to study the Rf values with different solvents in relation to the probable colours present. The Rf values are given in Table 4.1.

Table 4.1

RF Values of Permitted, Water-Soluble, Acidic Dyes

S. no.	*Name of Dye*	*Rf Values in Solvent Numbers* (i)	(ii)	(iia)	(iii)	(iv)	(v)
1.	Ponceau 4R	0.95	0.36	0.42	0.29	0.33	0.29
2.	Carmoisine	0.61	0.04	0.56	0.51	0.56	-
3.	Amaranth	0.77	0.06	0.20	0.24	0.19	0.20
4.	Erythrosine	0.23	0.00	0.70	1.00	1.00	0.38
5.	Fast Red E	0.45	0.00	0.60	0.45	0.54	0.60
6.	Sunset Yellow FCF	0.78	0.26	0.65	0.49	0.45	0.56
7.	Tartrazine	1.00	0.26	0.30	0.26	0.18	0.22
8.	Indigo Carmine	-	0.07	0.30	0.28	0.21	0.27

Table 4.2:
Colour Reaction of Water-Soluble, Acidic Dyes

S. No.	*Name of Dye*	*Shade*	*Conc. HCl*	*Conc. H_2SO_4*	*10% NaOH*	*Liquor NH_3*
1	*2*	*3*	*4*	*5*	*6*	*7*
1.	Ponceau 4 R	Red	Red	Violet red	Yellowish brown	Orange-red
2.	Carmoisine	Violet	Little chango	Deep violet	Red	Red
3.	Fast red E	Red	Slightly darker	Slightly darker	Violet	No change
4.	Erythrosine	Yellow-Red	Orange-yellow	Orange-yellow	No change	No change
5.	Amaranth	Violet	Slightly darker	Violet to brownish	Dull brownish to Orange-red	Little change
6.	Tartrazine	Yellow	Slightly darker	Slightly darker	No change	No change
7.	Sunset Yellow FCF	Orange	Slightly redder	Slightly redder	Brown	No change
8.	Indigo carmine	Blue	Slightly darker	Slightly Darker	Greenish yellow	Greenish blue

ESTIMATION OF SYNTHETIC FOOD COLOURS IN FOOD PRODUCTS

A. For Samples containing Single Colour

(a) Preparation of Standard Curve

(i) ***Stock solution*** 100 mg of each reference colour is accurately weighed and is dissolved in 0.1 N HCl in separate 100 ml volumetric flasks. The volume is made up with the 0.1 N HCl in each case.

(it) ***Working standard*** 0.25,0.5,0.75,1.0,1.25 and 1.5 ml of stock solution (i) above of each of the reference colours is pipetted into series of clean and dry 100 ml volumetric flasks and

diluted to volume with 0.1 N HCl. Optical density (Od) of each of the reference colours is determined at the respective wavelength of maximum absorption as given in Table 4.3. Od is determined using spectronic 20 photoelectric colourimeter. The standard curve is obtained for each colour by plotting optical density against concentration.

Table 4.3
Absorption Maxima of Permitted Food Colours

S. No 1	*Name of Colour* 2	*Absorption maxima (nm)* 3
1.	Amaranth	520
2.	Fast red E	508
3.	Carmoisine	516
4.	Ponceau 4R	507
5.	Erythrosine	527
6.	Green FCF	624
7.	Indigo carmine	609
8.	Green S	636
9.	Brilliant blue FCF	630
10.	Tartrazine	427
11.	Sunset yellow FCF	482

(b) ***Column chromatography*** A known weight of the sample (approx. 5 g) is transferred into a glass stoppered separating funnel (SF). The colour is extracted with 70% acetone, the acetone extract is extracted with petroleum ether (PE) (40-60°C) in order to remove carotenoids and other natural pigments, if any. This extraction is repeated till PE extract is colourless. The acetone extract containing only synthetic food colours is passed through a column (2.1 x 45 cms) containing AL_2O_3. The absorbed colour is eluted with 1% NH_3. The eluate is evaporated to dryness on a hot water bath, the residue is dissolved with 0.1 N HCl and transferred

quantitatively to 100 ml volumetric flask, and the volume made up with 0.1 N HCl. Optical density of the dye solution at the wave length of maximum absorption (tabulated above) is determined using the same spectronic 20 photoelectric colourimeter.

The dye concentration is determined from the standard curve.

B. For Samples containing mixture of colours

(i) ***Paper chromatography*** The colours present in the sample are extracted as described above under column chromatography. The volume of the purified dye solution is made up to a known volume with 5 ml of water.

An aliquot (approx. 0.5 to 1.0 ml) of the purified dye is spotted on Whatman No.l filter paper. The chromatogram is developed using solvent No. v detailed earlier. After drying, the coloured spots on the chromatogram are cut and eluted with 0.1 N HCl. A blank is prepared by cutting an equivalent strip from plain portions of the chromatogram and this too is eluted with 0.1 N HCl.

100 ml of volume of elute is made up with 0.1 N HCl and the dye content is determined as described earlier under column chromatography.

(ii) ***Thin layer chromatography*** (TLC) The colours are extracted as under column chromatography and the eluate is concentrated. The final 5 ml volume made as (i) (above).

Preparation of TLC plates 50 ml of starch solution (600 mg of soluble starch dispersed in 100 ml of glass distilled water heated to boiling to gelatinize starch) and 50 ml of 1.25% solution of Na_2EDTA is added to 50 gm silica gel (NCl Poona preparation) without any binder. The slurry is well mixed and then spread using applicator on glass plates 20 cm^2 to thickness of 0.5 nm. The plates are initially air dried and finally at 120°C for 2 hours.

An aliquot of the purified dye is spotted on TLC plates, the chromatogram is developed using amyl alcohol: glacial acetic acid : water :: 40:20:20 as the solvent system. The plate is removed and dried. The coloured spots on the plate are scrapped and transferred to test tubes. Colour is eluted using 0.1 N HCl. A blank is prepared by scrapping from the plain portions of the plate. The eluates are diluted to a known volume (say 100 ml) with 0.1 N HCl. The dye content is determined as described under column chromatography.

Oil Soluble Colours

Procedure

(i) ***Absorption of colours on Silica gel*** About 5 ml of sample is dissolved in 25 ml of hexane. About 10 g of silica gel of column chromatography grade and 2 g of anhydrous Na_2SO_4 are added to the solution. The contents are mixed by a magnetic stirrer for about 5 minutes. The supernatant liquid should be colourless, if not then a further quantity of silica gel 2.5 ± 0.5 g is added, mixed by stirring like before for 5 more minutes. Oils/fats get removed by repeated stirring 4-5 times with 25 ml of hexane and draining out each time.

(ii) ***Recovery of Colours from the gel*** The colouring matter absorbed during (i) is extracted from the silica gel with 2-3 volumes of di-ethyl ether 15-20 ml each. Total ethereal extract is evaporated in a porcelain dish on a hot water bath.

(iii) ***Separation of interfering matters from colours*** The residue from (ii) is dissolved in 0.5 ml of diethyl ether and applied as a band on a preparatory TLC plate of 300 micron thickness. The plate is developed in a hexane chamber and then air dried, the colour bands are scrapped into a small conical flask, 1 g of Na_2SO_4 (anhydrous) and 10 ml of diethyl ether are added. The contents are warmed gently for a while on a hot-water bath. The ethereal extract is decanted into a porcelain dish, extraction repeated for complete recovery of colour. All the decanted extracts are combined, and concentrated to about 0.5 ml.

(iv) ***Colours identification*** According to colour concentration 35 ± 5ml of the concentrated extract (iii) above, is spotted on silica gel plates of 300 micron thickness along with the reference oil soluble colours dissolved in diethyl ether. The plates are developed using any one of the following solvent systems.

(a) Benzene : Hexane : Acetone : Acetic Acid :: 40:45:7.5:1 (v/v).

(b) Benzene : Hexane : Acetic Acid :: 40:60:1 (v/v)

(c) Toluene : Hexane : Acetic Acid :: 50:50:1 (v/v).

The plates are air dried, Rf values of the coloured spots measured and compared with reference colours. Nevertheless, colour identification is confirmed by spraying reagent prepared by dissolving 2.5 g of boric acid and 1.5 ml of chloroacetic acid in 100 ml of HCl of sp. gravity 1.19. Change in colour is inferred as:

(i) Curcumin and turmeric spots change to saffron colour and red colour respectively.

(ii) Sudan TV (red) changes to violet colour.

(iii) Butter-yellow changes to brilliant pink colour.

(iv) Chlorophyll (bug green) changes to shining green and so on.

Even in the absence of standard colours, colours can be identified by using different solvent mixtures and subsequent spraying with appropriate reagents.

Chapter 5

Flavouring Agents

INTRODUCTION

Under Rule 63 of P.F.A. Rules 1955, flavouring agents include flavour substances, extracts/preparations, which are capable of imparting flavouring properties, namely taste or odour or both to food. Thus understanding the anatomy and physiology of taste and smell is more important than knowing the chemistry and physical properties of materials imparting the characteristics.

Classification of food on the basis of flavour has been probably first enunciated in the *Gita,* (xvii-8, 9 and 10). Stanza 8 groups together foods which are juicy, lubricated, well preserved and pleasing to mind are the best foods and liked by saintly persons. Stanza 9 groups foods on the basis of taste, viz. bitterness, sourness, saltiness, pungency and devoid of lubrication. Stanza 10 lists foods which are semicooked, dry, stale and putrefied. Thus odour and taste are the two main characteristics for judging the quality and acceptability of any food.

TASTE

There are four primary taste sensations, sweetness, bitterness, sourness and saltiness. All other tastes are permutations and combination of these. Mostly, bitterness is perceptible in the posterior region of the mouth and back of the tongue whereas sweetness is discernible on the tip of the tongue. The savoury food or ingredient must be in solution, only then the nerve endings on the tongue get stimulated and the sensation gets spread to the surroundings. A dry tongue has no role in taste perception. The sensitivity of taste buds is affected by the solution temperature, extreme in either way lower or annihilate the sensitivity of the nerve endings. 20 ± 10°C temperature range give optimum results. Physical/mechanical action within the buccal cavity is a prerequisite while judging a product organoleptically because it facilitates flavour penetration into the mucous membrane thereby enhancing the irritability of the sensing end organ.

Threshold is the point at which a stimulus to the sensory organism is just sufficiently intense to be felt. Taste threshold at the tip of the tongue is as little as 250 mg of salt or 500 mg of sugar dissolved in 100 ml of water at 20°C. Highly trained individuals have a even lower limit. Sides of the tongue are sensitive to 7 mg of HCl in 100 ml of water. The flavourist must be free of any limitation in detecting minute stimuli. For example, antihistamines, antibiotics, antimetabolites, anti-inflammatory as well as antithyroid drugs may all effect smell and taste, reports Schiffmann (1983). Elaboratively, a diabetic would not be a good judge of sweetness as the excess sugar in the blood may cause a sensation of sweetness from a substance that is devoid of sweetness, due to the false stimulation of the taste nerves. Likewise a jaundiced person may feel bitterness due to the stimulation of taste nerves from the bile in the blood. A heavy smoker and a *paan* eater are poor tasters because due to effect of lime in the *paan* the sensitivity of tongue diminishes considerably even to pungency of chillies and peppers.

Random House Encyclopedia states that many taste buds on the tongue decline up to 80% from the age of 20 to the age of 80. By the age of 60, 70% of taste acuity has been lost, therefore most panel members should be young and the rest a well disciplined panel of elders.

Saltiness is discernible on every part of the tongue.

Olfaction

The most acceptable theory suggests that a dissolved molecule of a specific size and shape is absorbed and penetrates the receptor membrane leaving a temporary hole. This permits local depolarization. The ensuing duration, size and rate correspond to the molecular characteristics of the dissolved molecule.

However, mechanical factors interfere with the access of the odour to the receptors. The simple airway obstruction gets complicated by drying up of the mucosa, alternatively by oedema. Blockage of the nasal package due to allergic rhinitis, Wegner's granulomatosis or mechanical destruction can occur. Similar effects occur due to deviated nasul septum, foreign bodies or polyps. Further, renal and hepatic failure, diabetes and influenza also damage or interfere or affect smell perception. Effect of antihistamines, etc. have been already stated above as per Schiffman (1983). Depressive or psychotic illness alters smell appreciation.

Odours are complex mixtures of different compounds, each one at a low concentration but studies in olfaction reveal that chemical constitution is related to olfactory stimulation. Usually, the olfactory potency of chemical compounds belonging to a homologue series increases progressively from the lowest member of the series to the highest. The odours of monoatomic alcohols, for example, increase in strength from -CH_3 through -C_2H_5, -C_3H_7 and so on, the relative potencies of CH_3 : C_5H_{11} :: 1 : 1000. Also compounds resembling in their physical and chemical properties tend to have odours possessing certain common characteristics. Nevertheless, the odour character due to osmorphic groups like -OH, C = O, -NO_2, etc. on the benzene ring is not dependent so much upon the particular radical present as upon the position which any of them occupies in the benzene ring. Strong absorption of infrared rays is a characteristic common to many odours. Small wonder if these two properties have a common bearing

Olfactory compounds meet the nasal mucosa and in order to produce a smell, need a high water as well as lipid solubility. Man discriminates a large number of different smells, and the olfactory mucosa and pathway are rapidly fatigued, although they recover quickly.

Sensory Adoption The olfactory receptors adapt rapidly. It is a common experience that a disagreeable odour which when first smelt is almost overpowering, soon becomes imperceptible. Though lost for one particular odour the sense of smell is retained for other odours. This phenomena is, therefore, not due to fatigue of the olfactory mechanism, but because of sensory adoption. The rate of adoption varies for different odours. Olfactory adoption begins to develop from the first smelling, the threshold rises gradually until complete insensitivity to that particular odour is reached. Even a previous period of exposure to a given odour raises the minimum concentration at which it is perceived for a considerable length of time afterwards. For example, the sensitivity of a person, who had been for a time in the operation theatre, to the odour of ether was below normal even after several hours.

Resuming from taste threshold, odour threshold concentration can vary by 10^{10} depending upon the chemical nature of the stimulus. The threshold of perception is lower than identification, i.e. an odour is sensed before it is recognised. Smell does not have an absolute threshold and the threshold corresponds to the level of inhibitory activity generated by the higher centres.

Olfactometry Man appears to be better at detecting the pleasantness of an odour than at identifying it. This is based on cultural factors. If two odours are mixed, the resulting intensity is always less than the sum of the two individually perceived intensities and is dominated by the stronger component. There is no interaction between the individual receptor cells connected to the olfactory bulb by non-myelinated nerve fibres.

Amoore (1969) has suggested about 30 primary odours for humans on the basis of sterio chemistry of compounds and the variation of anosmia to substances present in man. The human being has difficulty in detecting and identifying variation in intensity of more than 17 odours. Since the human beings do not rely on conscious detection of odour, only its quality, therefore, training is imperative for scientific experiments and for occupations which require a good nose and tongue. An obvious discriminatory mechanism has not been found in the nose at the receptor as well as in the olfactory bulb.

Olfactometry is a parameter of an individual threshold for smell. The unit is called olfacties. The Zwaarde Mcker's olfacto-meter method gives only approximate results. This has been improved by Elsberg and Levy and their method is known as Blast method, described below.

Blast Method 30 ml of an odorous liquid is placed in a bottle. A tube is connected to a double nose piece fitting in the nostrils. By means of a syringe connected to another tube and also connected to the bottle a measured volume of air is forced from the bottle in one blast at a constant pressure while the subject holds its breath. An equivalent volume of odour laden air is thus forced into the nose. The volume of injection is increased gradually in successive blasts until the odour is just perceived and can be named. The smallest volume necessary for identification is called the Minimum Volume Identifiable Odour (MVIO) or the olfactory coefficient.

Pertinently it is repeated that to determine accurately any particular odour, it is sniffed. This is to compress together the septum and outer wall of the nose at the front of the respiratory passages so as to divert the inspired (inhaled) air to the olfactory area. Diffusion is relatively slow process and is probably of minor importance in bringing the odorous material to the olfactory endings. Even though the nose is laden with odours air, one cannot smell with a choked nose because the breath is held up.

Organoleptic analysis of seasonings Sense of smell has a major bearing on appetite. Many texts suggest the testing of seasonings for aroma and flavour in a standard neutral soup base or white sauce. Since a neutral soup or white sauce base is not always of the same consistency, texture or colour, as the intended product and subsequently gives a different mouthfeel, which is a significant part of the overall flavour. Also, the base may selectively absorb portions of the seasoning and give a false inference. Therefore, only room temperature, distilled water whipped with air in the temperature range of 20 ± 10°C should be used for the initial comparative testing of spices and seasonings. Confirmatory tests of seasonings should always be made in the final product in which they are to be used.

Before any odour or taste evaluation of samples is done, the tester must ensure outside non interference lest it may detract from his/her subjective analysis described as follows.

1. The method of preparation of the sample should be as simple as possible and conducted in such a way that any diluents, materials or equipment used should not impart any foreign odours or flavours. Only tasteless and odourless distilled water should be used wherever possible. It may be desirable

to whip clean, fresh air into the water by means of a blender that has been cleaned and sterilized to remove any flatness inherent in the distilled water. Ordinary city tap water in many areas may contain high amount of chlorine, sulphurous compounds, or other contaminating substances that could effect delicate taste comparisons.

2. The level of dilution should neither be overpowering nor too weak to detect various finer odour nuances.
3. The remnant flavour of a tested sample must be completely rinsed out. Bottled carbonated water is recommended for this purpose.
4. Organolepticity should be compared/tested in an area devoid of any interfering odour. A small air-conditioned room, used solely for this purpose is ideal.
5. The testing room should have a uniform northern day light and no fluorescent light. In the absence of natural day light, lamps are available which duplicate northern light.
6. The flavour analyst should not have used any perfumed soap, cologne, after shave or perfume capable of interference with odour detection. Besides, there should not be any evidence of smoking in the room or on one's clothes, fingers, which may detract one's sensitivity.

Our knowledge of the chemical composition of flavours has increased at an exponential rate since early sixties by virtue of many innovations that have been taking place in the analysis of flavours in food matrices.

The concept of flavour as a product, distinctly separate from the food stuffs which contain it, is not a familiar one to the general public. Yet in the 20th century, a small but highly sophisticated industry has grown up around it. The initial impetus for the production of flavours came from chemical inventions like *vanaspati* which provided same food value as *ghee yet* much cheaper in comparison to *ghee.* As knowledge of the relationship between nutrition and health has evolved, the primary motivation for the use of these newer food sources is often health related rather than the economics. But in either case, there is no chance of public approval of these products if their flavour is not acceptable like margarine which could replace table butter. From these modest beginnings the flavour industry has grown to the point where its products are used in an amazingly broad spectrum of consumer products. Many products labelled 'all natural flavour' contain products from the flavour industry which have been compounded from such natural sources as essential oils, oleoresins, solid extracts and fermentation products.

The principal areas for which flavours are analysed are,

(i) quality control and assurance,
(ii) to identify new components found in the natural raw materials,
(iii) to determine the stability and the functional properties of flavour ingredients,
(iv) regulatory compliance and enforcement,
(v) to match existing flavours and food products and
(vi) to assist in developing new flavours and food concepts.

Consequently, in flavour industry the analytical procedures are diverse, so are the chemists employed and their responsibilities. The flavour chemist has to utilize his chemical knowledge as well as his sense of taste and smell in creating new flavour products.

Table 5.1: Functions of the Chemist in the Flavour Industry[1]

Title	*Description*
Research chemist	Develops innovative methods for the synthesis and analysis of flavor chemicals. Seeks to develop new concepts; applications and delivery systems for flavours.
Quality control chemist	Insures that all raw materials and finished products conform to specifications. To accomplish this task, the Quality Control Chemist must use organoleptic, chemical and instrumental methods of analysis.
Instrumentation chemist	Develops methods for the separation, identification and quantitation of both synthetic and natural flavour materials utilizing both chromatographic and spectrophotometric techniques.

Title	*Description*
Synthetic organic chemist	Designs and synthesizes molecules which will have applications to flavors. Synthesizes rare and expensive natural products from less expensive precursors.
Food chemist	Evaluates the compatibility of a flavor base with the foodstuff for which it is intended. Experiments with new food applications and processes where flavors are used.
Process development chemist	Supervises the scale-up of laboratory processes to commercial production.
Chemical engineer	Designs the equipment necessary for the large-scale manufacture of flavour products.
Flavour chemist	Blends natural and synthetic material from the approved list of raw materials (GRAS) to create new flavours. Although scientific information is utilized, a flavour chemist's work is based primarily on art and experience.

The modern analytical techniques used for the structural elucidation, separation and identification of complex organic, inorganic and biomolecules that are standard in all industries (pharmaceutical, chemical, polymer, etc.) like NMR, IR, UV, X-ray, AA, MS, GC, GLC, HPLC, HPTLC and the combined techniques of GC-MS, MS-MS, GC-FT/IR, or ancillary techniques of field desorption, chemical ionization, MS, etc. is beyond the pale of this monograph as these are so wide and already well explained in several books on these respective subjects. Recent works on these analytical systems has, however, cautioned these workers on the unguarded discrepancies[2] likely to creep in their final results.

After identifying each flavour chemical in the sample, the flavour chemist is confronted to determine the origin and significance of these chemicals which may vary widely. Clearly the task of obtaining a good flavour match is more than simply correctly identifying the peaks in a chromatogram.

Flavours are generally found dispersed in multiphase systems. While the flavour is very amenable to analysis by capillary GC/ MS the foodstuff in which it is incorporated definitely is not. Incomplete removal of such non-volatiles as aminoacids, carbohydrates and proteins will result in the formation of artifacts from pyrolysis on hot surfaces in the chromatograph, as well as rapid deterioration of the chromatographic systems's performance. The concentration of the individual flavour chemicals may greatly differ between the various phases present.

Apart from requiring maximum detector sensitivity, the very low concentration levels introduce discrimination effects into the quantities of flavour chemicals. While a few microlitres of oil are more than sufficient for GC/MS, experience has shown that, at this level, absorption on glass surfaces in the extraction vessel can lead to selective removal of certain constituents of the mixture. Further, many of the most potent flavour chemicals are sufficiently polar to show non-linear concentration curves even with modern fused silica capillary columns[3].

Usually low molecular weight volatiles (less than 400 Daltons) contribute flavour, but the following non-volatiles are generally their captors in bulk quantities.

Table 5.2
Predominantly Non-volatile Flavour Materials

No. Type	*Definition*	*Example*
1. Concrete	Extract of previously live plant tissue. Contains all hydrocarbons soluble matter and are usually solid, waxy substances.	All flower (Jasmine, Orange flower, Rose Carnation, etc.)
2. Absolute	An alcohol extract of a concrete which eliminates waxes, terpenes and sesquiterpenes	All flowers (Jasmine, Clove, Cubeb, Cumin, Lavender, Chamomile, etc.)

No. Type	*Definition*	*Example*
3. Balsam	A natural raw material which exudes from a tree or plant. They have a high content of benzoic acid, benzoates and cinnamates.	Peru, Tolu, Canada, Fir needle, etc.
4. Resin	They are either natural or prepared. Natural resins are exudations from trees or plants and are formed in nature by oxidation of terpenes. Prepared resins are oleoresins from which the essential oils has been removed.	Orris, Olibanum, Mastic, Cypress and flowers.
5. Oleoresin	Natural Oleoresins are exudations from plants, prepared oleoresins are liquid extracts of botanicals yielding the oleoresin upon evaporation.	Gurjum, Lovage, Onion, Pepper, Meliotus, etc.
6. Extracts	Concentrated products obtained from solvent treatment of a natural product	St. John's bread, Rhatany root, Mate Fenugreek, etc.

To detect these products in flavours, it is pertinent to analyse for these non-volatiles. This is altogether a different methodology vis a vis flavour volatiles.

Many peaks detected in the chromatogram are artifacts because flavour contains many chemicals capable of reacting with other components present within the flavour, the food base or the container. Reactions like oxidation, acetal formation, hydrolysis, ester exchange, transesterification and Schiff's Base formation produce artifacts and should not be mistaken as a characteristic of true flavour.

Paucity of published data about the spectroscopy or chromatography of these chemicals adds to aforesaid mistaken identity. However, these hurdles have been removed in the present day flavour analysis as summarized in the following paragraphs.

Non volatiles The analysis of non-volatile fraction of flavour is important from two different angles.

(i) Several natural products compounded into a flavour contain the organoleptically active volatiles in micro quantities hence their detection is practically possible through the identification of the more abundant components like sugars, OH-acids, aminoacids, flavonoids, etc.

(ii) The taste sensations described earlier and tactile sensations as heat and texture are produced primarily by non-volatiles.

Their analysis is commenced by separating the complex mixture into pure substances which can be examined spectroscopically. Separation is effected by HPLC using normal phase, reverse phase or ionic phase separation with attending modifications. Absence of general methodology does not identify in a single analysis any major non-volatiles as is done by GC/MS for the volatiles because

(i) Liquid chromatography (LC) lacks the separating capability of GC. A good LC column has 10,000 theoretical plates, whereas a modern capillary GC column has 2,50,000 plates as stated by Jennings *loc cit* (pp. 2-13).

(ii) The carrier in LC cannot be removed like in GC, thereby taking away the most powerful spectroscopic identification tools like MS, NMR and FTIR. While certain LC/ MS techniques are used, their spectra are not comparable to the published ones in the chemical literature or accumulated in individual laboratory data bases.

(iii) Relative retention indices have not been used as successfully as in GC. This is partially due to the lack of a convenient retention standard like Kovats which is based on the retention behaviour of homologues in GC. Besides, the necessity for several stationary phases and carrier combinations requires several retention data determinations for each chemical.

LC provides the greatest flexibility to choose detector, refractive index and short wave length. UV detection yields universal detector response. Fluorescence, pulsed amperometric and suppressed conductivity can be used for selective detection. This detector selectivity can be used to compensate for limitations in LC separation capacity. Current LC technology necessitates extensive method development work for each class of substances to be analysed.

The type of flavouring material determines their specific LC procedures. The alkaloids present in fenugreek, quebracho and passion flower may serve as markers for their identification. Vanilla products contain unique hydroxy acids detectable by UV absorbance at 210 nm and reversed phase chromatography. Nevertheless, several more exotic flavouring materials lack reliable methods for their detection in a finished product*.

Volatiles Since the first commercial instruments became available gas chromatography has been the most important tool for the analysis of flavours. Because olfactory sensation is such an important aspect of flavour, the separation and identification of the extremely complex mixture of volatile organics present in both natural and artificial flavours is of primary importance. Since these mixtures may contain hundreds of substances, the high resolving power of capillary columns is an absolute necessity. The availability of capillary columns with cross linked stationary phases has greatly enhanced their utility to the flavour analyst. Cross-linked columns make practical the use of sampling techniques like purge and trap, head space analysis and splitless injection which often cause severe column unloading. Even polar cross-linked columns, such as PEG types, can tolerate amounts of water which would quickly destroy conventional columns of this type. The analyst can experiment freely for purge flow rate, purge time, sample temperature and volume, without damaging such costly columns. Cross linked capillary columns yield numerically more theoretical plates, lower bleed rates and higher maximum usable temperatures and markedly better reproducibility of retention indices than do conventionally coated capillary columns; splitless injection usually introduces ml amounts of the organic solvent to extract trace organics from an aqueous medium. Thus losses of extremely volatile, easily absorbed or thermolabile components can be avoided during the solvent

* Nadim A. Shaath *et. al., proceedings of the 5th International Flavour Conference,* July 1987, Greece.

removal operation. Analysis of minor components of essential oils is also useful.

The first part of flavour analysis is that peaks separated by the capillary GLC be qualitatively identified. MS of the unknown peak is matched/compared with stored spectra, keeping the following points in mind.

(i) The published/stored spectra may not be having samples spectra, may not be authentic due to problems of sample purity and instrumental differences. The spectral library must be assembled stepwise from the world's chemical literature, cross-checking various published sources of spectral and compositional data with the laboratory's own data base.

(ii) Many closely related terpenoid compounds yield identical MS making identification difficult if not impossible, therefore, the retention index must always be checked to confirm the MS identification. This has been made a standard practice in the flavour analysis to connect a non-polar (CH_3-silicone) and a polar (PEG-20M) 50 meter capillary column via two-hole ferrule to a common injection part. The resulting simultaneous chromatograms are then cross-compared by area comparison and finally matched with the GC/MS chromatogram which is usually run on a CH_3-silicone capillary column. The combination of polar and non-polar retention indices, together with MS, generally yields a satisfactory identification.

Despite identification of all peaks obtained from a GC-MS analysis, duplication or reproduction of precisely the same organoleptic properties of the studied sample is not attained by combining the identified components. The reason is the extremely low perception threshold of many flavour materials. Such compounds are identified by,

(i) preliminary fractionation of the flavour sample prior to GC-MS. It involves column chromatography, fractional distillation or selective chemical extractions. But these procedures are time consuming, difficult to quantitate and prone to formation of artifacts.

(ii) Use of selective detectors capable of 'seeing' the powerful odourants but 'blind' to the bulk of volatiles. This is rapid because further manipulation is dispensed with.

Usually used selective detector is the Selected Ion Monitoring (SIM)[4]. Since any fragment ion abundant in the material under analysis may be used, the sensitivity and selectivity can be greatly enhanced for any target compound. If the target compound has abundant ions at usual mass numbers, the potential enhancement is still greater. Since the number of ions that can be monitored per run is limited, it is essential to choose as targets those substances which can be significant organoleptically at trace levels. Essential oils which are used at very low levels as fortifiers in natural flavours can often be detected in this fashion. By setting the MS to monitor only these masses and calculating the retention indices of the resulting total ion chromatogram, the identity and quantity of the test material at the low levels can be positively achieved. Sandalwood oil or cedarwood oil are made up of very complicated mixtures. The former has a particular m/e 94 and the other 199 which is common to many components.

Pyrazines, pyridines, thiazoles and aliphatic amines, all nitrogen bearing compounds, comprise another class of low level but highly important compounds. Direct splitless injection of ml amounts onto cross-linked capillary columns—permits the Thermo Ionic Detector (TID), i.e. AFID alkali flame ionisation detector, to respond to N-containing peaks at ppm levels. Non-nitrogenous compounds are inert to the TID, except at very high concentration[5]. These are readily differentiated from the N-compounds by their typical over loaded shape of the peak.

The fact that a substance responds to the TID combined with its retention index on both polar and non-polar columns permits a reliable identification at levels well below those of conventional GC-MS without preliminary concentrations.

Sulphur compounds which may be repulsive in odour above trace level play an important part in food flavours. The flame photometric detector is suggested for their detection as done with (TID).

INSTRUMENTAL TECHNIQUES IN FLAVOUR ANALYSIS

The advent of modern analytical techniques like GC, UV, IR, NMR spectroscopy and Mass spectroscopy has revolutionised our understanding of the physiology/perception of flavour and aroma particularly the importance of minor components in them during recent years. As a result manipulation of naturals and synthetics has advanced a lot in perfumery industry but in food industry it has not gone far. A commercial flavour is usually a mixture of aroma chemicals, essential oils, gums, resins, emulsifiers, thickening agents, etc. The actual flavour ingredients will be about 1% .

Instrumental analysis involves

(i) isolation, concentration and separation,

(ii) detection and identification.

Water steam distillation, absorption, CO_2 reduced pressure distillation, expression, formation of derivatives, head space vapour collection, high vacuum degassing, solvent partition, vacuum sublimation are some of the generally used isolation techniques.

Continuous steam solvent extraction, its modified form and liquid-liquid extraction are employed for extraction of volatiles.

Adsorption/desorption, displacement, solvent recovery, freeze concentration (e.g. votator continuous, freeze concentration system), molecular distillation, vacuum distillation/fractional and zone concentration are some of the widely used concentration techniques.

Chromatography can be used as an analytical and separation technique and moreover can form the basis for most of the sophisticated instruments employed for flavour analysis. The use of silica gel impregnated plates enable us to separate hydrocarbons which are otherwise difficult to be separated. Column chromatographic separation has an edge over fractional distillation because no heat is applied to flavour components during separation. Narayanan *et. al. (loc-cit)*[5] have found it useful for separating clove oil components.

In gas chromatography, due to difference in adsorption coefficient in the mobile phase, enables the separation of components while difference in partition coefficient in the stationary phase forms the basis of separation. The technique is generally used for the analysis of gases and liquids with boiling points up to 350°C. The analysis is usually carried out at temperatures below the boiling point of the sample.

Elution technique is the commonly used method in GC even though frontal and displacement analysis can be adopted in specific cases.

Gas liquid chromatography, also called vapour phase chromatography, is the most popular and useful of all chromatographic techniques.

Chromatographic technique of identification of a component is the widely used method in flavour/perfumery industry. GC identification is done by noting its retention time which is a fundamental property of the compound under particular conditions used.

Comparison of retention time of an unknown compound with that of an authentic compound is done by coinjection and peak enhancement method. Relative retention time which is more reliable than absolute retention time is measured relative to a standard. This minimises the errors due to small changes in column temperature or flow rate. The most widely used system for qualitative analysis is the 'Retention Index' (RI) proposed by Wehrli and Kovats. The techniques of functional group analysis provides another important method for the identification of functional groups in a compound.

Mass Spectrometry (MS) Produces a beam of gaseous ions from a compound, separates the ions according to the mass to charge (m/e) ratios by the action of electric and magnetic fields and records their masses and relative abundances. Under low pressure (10^{-6}) mm Hg) a molecule when bombarded with an electron produces an ion which may again fragment into smaller ions. Usually the positive ions thus produced are analysed since their production is rather easy. The MS consists of an ion source, mass analyser, ion detector and vacuum system. Usually the electron impact method is used for producing ions (electron energy (eV) range 20-27) and is satisfactory in most cases. Chemical ionisation may be more suitable for mixtures of less stable compounds. MS of a compound informs about its molecular weight, molecular formula, nature of functional groups present and presence of certain elements.

GC-MS analysis This couple is a powerful analytical tool for qualitative as well as quantitative analysis of flavours, even in micro-quantities. But the two systems are inherently not compatible. GC operates under pressure and MS under vacuum. There is a 10^2 to 10^4 dilutions of the sample with the carrier gas hence an interface is applied to concentrate the sample molecules by excluding the carrier gas. The interface may be (a) effluent splitters, (b) molecular separators and (c) direct coupling. Jet separator is the common interface of most commercial GC-MS systems.

Computerised data processing has considerably reduced the involved labour. FT-IR employs an interferometer and computer technology to measure precisely, rapidly and with greater sensitivity the spectrum of a compound. Conventional dispersive IR requires usually 15 minutes/scan at 8 cm' per resolution to cover the mid-IR range. An FT-IR spectrometer operating at 0.6 sec/scan hence can accomplish 1500 scans in 15 minutes, resulting in an approximately 30 times increase in signal/noise and a significant decrease in the amount of sample required.

By coupling GC-FT-IR for on line measurement of the spectra of GC eluates. The spectral frequencies given by the instrument in the absorbance mode are so precise that the final spectrum is completely amenable to sophisticated computerised search of reference spectra libraries. Sample requirement is also minimised (10-100 ng).

Instrumental identification

1. Infrared absorption of a compound is dependent on its chemical nature and is highly characteristic. It can be used for detection of functional groups, study of cis-trans isomerism, etc. Infrared spectrum of an essential oil as such gives a lot of information to an experienced chemist regarding its purity and can be deemed peculiar to each oil.
2. Nuclear Magnetic Resonance (NMR) depends on the separation and summation of the resonance signals emitted by protons in different environments. For example, the protons

in a terpene may have different signals for protons which differ in electronic environment. It is possible to assign the signals to methyl, methyl-on double bond, gem dimethyl groups, ethylenic protons, carboxyl methyl, etc. present in a terpene.

^{1}HNMR gives details about the pattern of hydrogen atoms attached to the respective carbons and their spatial arrangements.

^{13}C NMR also gives concrete ideas about carbon skeleton of the organic molecules and more information about quarternary C-atoms, double bonds and ketonic groups.

Fourier Transfor-NMR (FTNMR) - a multinuclear instrument, so apart from ^{13}C and ^{1}H, one can obtain NMR data on nearby 50 nuclei. Here the sample is irradiated with an intense pulse of RF energy for a short time and the resulting FID signals are collected, digitated and stored in a computer. Several low-cost NMR spectrometers are available.

Proton Magnetic Resonance Spectra (PMRS) of all the essential oils are recorded on a 60 MHz Hitachi R_{248} high resolution NMR spectrometer with TMS as internal standard. Solutions are made with 50 ml of an essential oil and 0.3 ml of CCl_4. The components are identified by the position and splitting pattern of the chemical shifts by comparison with the data available in literature. Integration is not taken into consideration for identification. Chemical shifts are given in d (delta) values (ppm).

Rao[6] has reported the salient features, of the PMRS of several essential oils as well as the corresponding identified components. He concludes that an essential oil can be certainly identified, at least tentatively by its PMRS and in most of the oils a direct structural identity of the major components can also be obtained.

As stated earlier most of the components in a mixture are analysed automatically either by GC or GC-MS by comparing with previously compiled data in libraries. In spite of its speed and ease of use it should be used with great care as has been pointed out by the working group on Methods of Analysis of the International Office of the flavour industry (IOFI) in a recent publication on this subject. In their final conclusion these experts declare:

"Neither GC nor MS evidence on its own can be satisfactory basis for the conclusive evidence. Even a combination of two may still require an additional form of analysis, such as NMR or IR spectroscopy in order to ensure absolute certainty"[7].

Certain types of highly polar components generate unreliable chromatographic retention data and situation becomes worst with aging columns. No computerised library search has any power of decision because the final decision belongs to the analyst. However, data compiled in search libraries is recorded under a variety of conditions using different instruments and parameter settings*.

"There are more than 230 naturally occurring sesquiterpenes with molecular weight 204. These are expected to be eluted in GC on both polar and non polar columns within a range of about 300 kovats units and one can expect them to display similar mass spectra (m/2 = 105,119,133,166...). Therefore, it is quite understandable that GC/MS identification alone can be hazardous".

Recent compilation of published GC retention data of sesquiterpenes are also not reliable .

Identification of non-isolated constituents of essential oils can be considered feasible when MS and GC retention indices are strictly identical to those of reference samples present in data bases developed by the analyst using the same methodology. The analyst should check the published data and discard non confirmed information.

Another source of contamination may be from the analyst's own skin[9].

Chemically generated artifacts from isomerisation, cyclization, racemisation, hydrolysis, elimination and interesterification are well known hence need to be minimised when performing analysis. The classical parameters such as sp. gravity, optical solation, refractive index, should be measured after proper separation using GC/HPLC. The analyst should use data reported by chromatographic detection and integrators after ascertaining its reproducibility, repeatability

* Gurdeep Singh, I.P. Singh Kapoor and S. Modi Mannon, *Indian Perfumer,* 1995, 39(4), 149-153.

and accuracy. Papas *et. al.* have reported that when four GC integrators were tested using simulated data, they produced not only significantly discrepant results between instruments but also, for a given instrument; wide effects of noise and tailing on peaks had different heights and widths[10].

Gurdip Singh *et. al. (loc.cit)* attribute this due to pressure/flow disturbances inside the injector or in some other part of the sample path or in data processing. *(Standard samples do not necessarily produce standard chromatograms.)*

Notwithstanding, modern analytical techniques have reached a high degree of sophistication and sensitivity that a tiny amount of any component in a mixture can be detected. The increased resolution and specificity are expected to find new applications in flavour analysis and complex mixtures of natural products chemistry. The need of the day is that analysts should interpret analytical information responsibly, pursue their work to build accurate, reliable and large scale data bases and find accurate protocols.

Electronic nose Electronic nose is a hybrid system which uses a combination of different sensor technologies like MOS (metal oxide semiconductors), conducting polymers and quartz microbalance, etc. for taking digital finger prints for aroma, flavour, odour, etc. It is manufactured by Alpha M.O.S. S.A., 20 Avenue Didier Daurat, 31400 Toulose-France.

Detection principle

(i) The basic detection system is made of six small semi conducting odour sensors and can be upgraded with several more arrays, of six elements (for a total of 6, 12 or 18 sensors). Metal oxide sensors (M.O.S.), conducting polymers or surface acoustic wave (SAW) arrays will complete the available range.

(ii) M.O.S. have different partial selectivities and respond to a wide range of odours. The sensors can easily be modified to suit your applications by selecting the devices from a current range of more than 40 different elements (SnO_2, ZnO, WO_3...).

(iii) The detection is based on a change in the electrical resistance of the sensors when the odour is present. Using multiple sensors in place of our alfactory receptors, the electronic nose mimics the human sense of smell. Neural networks help to classify new flavours according to odour descriptors.

FOX 2000 system It is an automatic calibration that has been implemented when the array of sensors is modified and each element can be turned to an optimised temperature. The unit also includes a temperature .sensor, a humidity sensor and a variable flowrate.

One measurement takes around two minutes for data acquisition and three minutes for base line recovery of sensors. Each sensor can easily be exchanged and a cleaning procedure is available to recover quickly the baselines. Alternatively, the instrument which is a dynamic system can be used in combination with or special measurement chamber connected to the inlet or with another flow injection system (head space, purge and trap ...).

Applications Electronic nose technology is useful as a rapid tool for verification of raw materials and permits production quality control before costly mistakes can happen. This is the desired objective of hazard analysis critical control point system, presently so much being persued by all industries including food industries. It is also a useful tool for research and development and to improve quality control and reduce production costs for example, the following industries are currently using electronic nose technology.

Food and Beverage Products

Food

(i) Inspection and grading quality of food by odour.

(ii) Meat: degree of spoilage, rancidity, boar taint detection in raw bacon.

(iii) Fish freshness and classification of sea food.

(iv) Grains and cereals: storage conditions, quality, contamination, adulteration.

(v) Dairy: milk quality, cheese maturation, butter varieties, yeast analysis.

(vi) Biscuits, crackers and cookies; control and monitoring of the cooking or baking processes,

(vii) Vegetable oils, flavour oils, synthetic oils-detection of adulteration.

(viii) Potato products: variation in flavours, frozen French fries, potato chip rancidity,

(ix) Cocoa beans: quality control, classification of chocolate, coffee and tea by origin.

(x) Shelf life stability.

(xi) Fermentation control.

(xii) Permeation and flavour scalping due to packaging deficiency.

Beverages

(i) Discrimination and authenticity of whisky brands; differences in whisky blending processes.

(ii) Carbonated beverages and coca drinks: pet bottles, natural flavours, blend, storage stress, counterfeiting.

(iii) Alcohol beverages: wine and champagne analysis, TCA in cork, beer quality and stability of grains (rice, hops, malts...).

(iv) Fruit juice adulteration, origin, blending and formulation, rancidity, grape fruit juice taint, packaging.

(v) Beverage container inspection.

DIRECT ANALYSIS OF FOOD AROMAS WITH ELECTRONIC NOSE

Quality control of flavours and aromas by organoleptic scoring is a highly specialised art and depends upon the skill of the expert. To acquire subjective quality control expertise it takes many years. Therefore, it is advantageous to have a method for food scientists to measure the organoleptic parameters of a product. Gas chromatography (GC) is one of the most common method to separate all molecules of the samples headspace and allows the measurement of the volatile compounds. But there is a need for a fast method capable of evaluating over all aroma intensity and quality.

Electronic noses, with multisensors and pattern recognition techniques, provide a method to measure odour quality in a very rapid manner after initial training phase given by the sensory panel or the analysts. They allow a finger print based on the sensor responses to be recognised as a certain quality of spice or contaminant.

Spices have always been very important all over the world through out time. They are very expensive constituents of food, for example, cardamom, cinnamon, saffron, etc. However there is a large range of different qualities depending up to their origins, the extraction process and also because of a large number of frauds.

Equipment The Alpha M.O.S. Electronic Nose system, the Fox 2000 consists of three main modules,

(i) A sensor chamber for measuring the characteristics of the odour sample, the temperature and relative humidity of the chamber. The basic system has a sensor array of six metal oxide sensors covering a wide spectrum of molecules that can be present in the head space of samples.

(ii) A sampling chamber containing the samples which can be heated upto 200°C with an accuracy of ± 0.1 °C. This chamber allows a generation of a reproducible headspace by stabilizing its temperature. The sampling chamber can handle large volumes of liquid and solid samples (60 ml to 500 ml.)

(iii) Software mat allows data logging of the sensor's behaviour. Linear and non-linear statistical data treatment are part of the software package and can give an on-line or an off-line description of the sample.

A. Theory of Operation

Metal oxide based sensors Metal oxide is a semiconductor material and is gas-sensitive. Oxygen in the air reacts with lattice oxygen vacancies of the bulk material and removes electrons from the conducting band.

$$n + \tfrac{1}{2}O_2 \text{ — } Os$$

The oxidation state is temperature dependent but O negative species occur at around 400°C. In the presence of a gas or a fragrant molecule R, the chemisorbed oxygen species react irreversibly and the products are usually evolved (e.g. CO_2 and H_2O).

$$R(g) + O(s)^- \rightarrow RO(g) + n$$

The resistance of the sensor thus decreases in the presence of an odour with the size of the response depending upon the nature of the detected molecules and the type of metal oxide. The response time depends upon the reaction kinetics, the head space and the volume of the measured head space, the flow rate of the carrier gas. It usually takes between 10 to 120 seconds to obtain the final response of the sensors.

The Electronic nose fox 2000 has an array of 6 metal oxide sensors which can be upgraded to 18. The flow-rate can be set between 1 ml/mn and 500 ml/mn.

Metal oxide sensors have a very good sensitivity (ppm or even ppb level for special sensors) for a very broad range of chemical compounds. Due to their poor selectivity (all sensors can respond to a single volatile compound but in different magnitudes), only the use of arrays leads to an induced selectivity (e.g. 'pattern of 6 sensors'). These sensors are made of a thin layer (50 mm) with an oxide film deposited on a ceramic file.

Sensors are usually heated between 175°C to 425°C. Selectivity of some sensors can be related to different catalytic amounts of the doping metal (Pd. or Pt. for tin oxide sensors) on the sensor surface. Time response also depends on temperature, on the type of gas and flow rate as well as on the average time to displace a head space of 60 ml which is about 10 to 20 seconds. Other metal oxides (both n-and p-type, ZnO, WO_3. . . $)_2$ are also available for this instrument.

The set-up of the fox system can be seen in Fig. 5.1

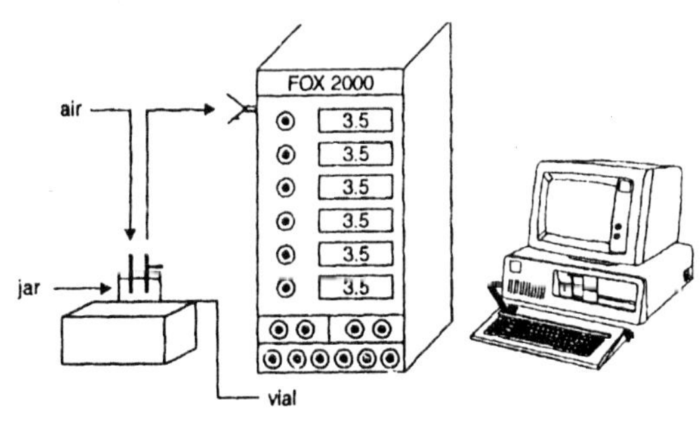

FOX System Representation

Figure 5.1

B. Operating conditions

The different spices are weighed in a microvial which is inserted inside a large 120 ml vial. The operating conditions are,

(i)	Carrier gas	Ambient Air
(ii)	Flow rate	320 ml/min
(iii)	Acquisition time	45 seconds
(iv)	Headspace generation time	30 seconds.

C. Sample preparation

(i)	Volume/Weight of sample	50 mg
(ii)	Measurement Cell Volume	120 ml
(iii)	Measurement Chamber temperature	30°C

D. Sensors used

(i) P 30/1: Polar Compounds (Alcohol..., etc.)
(ii) P 10/2: Apolar compounds (Hydrocarbons ...)
(iii) P 10/1: Similar to P10/2 but with selectivity.
(iv) P 40/1: Fluorinated compounds
(v) PA1: Food compounds
(vi) T 70/1: High sensitivity to compounds emitted during cooking process.

The analysis are made on a given selection of sensors, the sensor array is not optimised for this study. Better discrimination can be obtained if different sensors are used in this study. There are 42 different types of sensors which can be used to improve the discrimination obtained here.

E. Operating mode

After headspace generation time of 30 seconds at 30°C, the vial is connected to the inlet of the instrument. After an automatic or a manual injection, the vapours induce a decrease of the sensors resistance. The variation (sensor response intensity) is recorded as a function of time. Between each measurement, a stabilization of the sensors is necessary. Further experiments are carried out, only when the sensors get stabilized to their initial values.

The choice of analytical parameters (time, flow rate, temperature) and the sampling method are fixed so that all the headspace is displaced. The acquisition time can be reduced by increasing the flow rate.

Interpretation of Results

In order to obtain the qualitative and quantitative differences between the several product classes, graphical representations and statistical methods are used.

The star plot representation enables a comparison and visualisation of the qualitative (if normalised data is taken) or quantitative (if raw data is taken), finger print of the samples. Data at two different time points per sensor are plotted. Each data point is plotted on a different axis. The scale of representation is based on the range of responses obtained.

Statistical analysis such as cluster analysis, principal components analysis or discriminant analysis are used to determine the groups within the samples or to obtain the olfactory territory of a sample series and to study the product evolution during a process.

Summing up, Electronic nose's concept as an analytical instrument can be a perfect complementary tool to GC or to sensory panels to assess extremely quickly the quality of the spices or the kind of contamination presented by the samples with minimum preparation time. The Fox 2000 can be readily adopted for direct quantification and identification of the samples. It can made fully automated by connecting it to a classical dynamic or static GC headspace samples.

Besides the above objective analysis, a food technologist must also be a good subjective analyst. He should have a sensitive palate, a knowing nose, a perceptive eye, an economy oriented mind, an inspirational intuition and understanding of consumer preferences, a knowledge of marketplace and the ability to communicate clearly. This shall help the technologist to find out or develop cheaper substitutes without sacrificing quality in the final product. Each product a technologist uses or develops has a price label. A few paise saved by intelligent substitution saves millions at the close of an accounting year or shall enable the manufacture to price it competitively provided it is equal if not better than the competition. Technologists should be able to sit down with the marketing sales and cost accounting personnel to discuss the over all programme before, during and after development of a new product or the duplication of an existing competitive one.

An experienced, intelligent and seasoned food technologist working in the industry, is expected to be able to breakdown a food product or seasoning, primarily by observation and sensory analysis, and duplicate it within a few hours without the use of sophisticated laboratory techniques detailed earlier in this chapter.

He shall always keep in mind the ingredient costs and government restrictions on their use in food products as well as their functional characteristics.

The analyst must be totally familiar with the characteristics of each spice in its whole and powdered form, spice extracts in their fresh form, and record their attributes mentally. This is to be done repeatedly till he can identify them precisely blindfolded, tastewise and smellwise. Each spice extract is diluted by blending it on dextrose and the smelling and tasting exercise is repeated as described earlier till each one is mastered. Remember no extractive, whatsoever, is tasted at full strength.

To obtain the full flavour and aroma of a spice, Kenneth T Fanell suggests the following procedure.

2.835 g of the freshly ground spice in a 100 ml volumetric flask is filled to the mark with hot distilled water (98°C) cooled down to about 74°C within ten minutes, poured into 200 ml beaker or a brandy sniffer and the emanating volatiles are smelled with rapt attention to note their initial aroma and the succeeding nuances. Water colouration, if any, also to be recorded. Due to further cooling when the contents are at about 30°C, resmell and then taste. All the observations are carefully and attentively recorded. This exercise is repeated until your findings concur within themselves. Your descriptions may not always agree with other investigators, but they are your own and you are the one that will be using them later in your work.

Spice extractive is weighed out exactly the gram equivalent of one ounce of spice as tabulated in Table 5.3 (pp.150-58) blended on sufficient dextrose or non-iodized salt to make 100 g of the soluble spice. Blending must be complete and uniform. Exact 10 g = 0.1 oz of spice extractive is transferred to a 100 ml volumetric flask, volume made up with hot distilled water at about 74°C, contents shaken for complete solution. Sniffing and tasting is done like before for fresh spices. Accurate weighings yield favourably similar flavour as of fresh spice due to similar concentrations. Hotter water used for the fresh spice was to extract the entire volatile matter. If the sensory characteristics differ then the spice source whence the manufacturer's extract was made is scrutinized and compared for similarity with the employed extract as well as its equivalency with the reference table.

If the solvent odour is detected in the spice extractive then the original container is smelt immediately after opening it. If a solvent residue is present, its odour will be quite pronounced. Extractives, i.e. oleoresins must be practically free from solvents, less than 30 ppm. If more than this, the material and its source should be changed because solvent tainted extractives ruin a finished product.

After complete familiarity with all the sensory characteristics of the spices and their derivatives it is not difficult to analyse or synthesize the desired seasonings provided minute differences in sweetness, saltiness, bitterness and sourness are differentiated minutely and with rapt concentration. More dilute the solution more subtle is the perceived differentiation, therefore, quantitative solutions of the extract should be made in water at concentrations of 100, 50, 20, 10, 5 and 2 ppm and from these the lowest level of perception be made. A good taster should be able to perceive 50 ppm or less of each. Non-iodized edible salt, hydrochloric acid, sugar and quinine may be employed for saltiness, sourness, sweetness and bitterness respectively.

DECODING FLAVOUR BLENDS

Part VII of PFA Rules, 1955 pertains to packing and labelling of food. Under Rule 32 (a) the name/description of food contained in the package, and as per Rule 32(b) the names of ingredients used in the product in the descending order of their composition by weight or volume as the case may be, shall be specified on every package. However, in the case of artificial flavouring substances, the label may not declare the chemical names of the flavours, but in the case of natural flavouring substances or nature identical flavouring substances as specified under rule 63 (a, b, c) the common names of flavours shall be mentioned on the package.

Table 5.3 Spice Extractives/Equivalencies to Fresh Spices/Soluble Spices[a]

Extractive	Type[b]	g/Type equivalent to 1 oz ground spice	Extractives, on soluble, dry, edible carrier (%)	Minimum vol. oil in extractive vol/wt. (%)	Federal specifications: Extract on dry carrier (%)	Federal specifications: Vol. oil in extract vol/wt. (%)
1	2	3	4	5	6	7
Allspice (pimenta)						
leaf	E.O.	0.710	2.504	100	-	-
berry	O.R.	1.418	5.000	40	-	-
berry	S.R.	1.276	4.500	60	4.5	40
Anise seed	E.O.	0.710	2.504	100	-	-
	O.R.	2.339	8.250	15	-	-
Bay leaf (laurel)	E.O.	0.284	1.000	100	-	-
	O.R.	1.418	5.000	15	-	-
	O.R.	0.851	3.000	28	-	-

Contd . . .

Table 5.3 Contd . . .

Extractive	Type[a]	g/Type equivalent to 1 oz ground spice	Extractives, on soluble, dry, edible carrier (%)	Minimum vol. oil in extractive vol/wt. (%)	Federal specifications: Extract on dry carrier (%)	Federal specifications: Vol. oil in extract vol/wt. (%)
1	2	3	4	5	6	7
Basil	S.R.	0.213	0.750	40	2.0	40
Capsicum (red pepper)						
1,000,000 Scoville H.U.	O.R.	1.701	6.000	0	–	–
480,000 Scoville H.U.	O.R.	1.701	6.000	0	5.0	0
Caraway seed	E.O.	0.710	2.504	100	–	–
	S.R.	1.418	5.000	60	5.0	60
Cardamom	E.O.	0.851	3.000	100	–	–
	S.R.	1.134	4.000	70	4.0	50

Contd . . .

Table 5.3 Contd . . .

Extractive	*Type*[b]	*g/Type equivalent to 1 oz ground spice*	*Extractives, on soluble, dry, edible carrier (%)*	*Minimum vol. oil in extractive vol/wt. (%)*	*Federal specifications* Extract on dry carrier (%)	*Vol. oil in extract vol/wt. (%)*
1	*2*	*3*	*4*	*5*	*6*	*7*
Cassia	E.O.	0.430	1.517	100	–	–
Celery seed	E.O.	0.567	2.000	100	–	–
	O.R.	1.347	4.750	9	–	–
	S.R.	0.851	3.000	12	3.0	10
Cinnamon	S.R.	0.567	2.000	65	6.0	50
Clove	E.O.	4.260	15.026	100	–	–
	S.R.	1.701	6.000	70	6.0	70
Coriander	E.O.	0.142	5.000	100	–	–
	S.R.	0.851	3.000	40	–	–
Cumin	S.R.	1.418	5.000	60	–	–

Contd . . .

Table 5.3 Contd . . .

Extractive	Type[b]	g/Type equivalent to 1 oz ground spice	Extractives, on soluble, dry, edible carrier (%)	Minimum vol. oil in extractive vol/wt. (%)	Federal specifications: Extract on dry carrier (%)	Federal specifications: Vol. oil in extract vol/wt. (%)
1	2	3	4	5	6	7
Dillseed	E.O.	1.142	4.028	100	–	–
	O.R.	0.418	5.000	10	–	–
	S.R.	1.418	5.000	70	5.0	60
Dillweed	E.O.	0.284	1.000	100	–	–
Fennel	E.O.	1.418	5.000	100	–	–
	S.R.	1.843	6.500	50	5.0	50
Garlic oil	0.142	0.500	100	–	–	
Garlic (special)[c]	O.R.	0.567	2.000[d]	5g	–	–
	O.R.	2.268	8.000[e]	5g	–	–
Ginger, African	O.R.	1.134	4.000	28	4.0	25
Cochin	O.R.	1.134	4.000	28	4.0	25

Contd . . .

Table 5.3 Contd . . .

Extractive	*Type* [b]	*g/Type equivalent to 1 oz ground spice*	*Extractives, on soluble, dry, edible carrier (%)*	*Minimum vol. oil in extractive vol/wt. (%)*	*Federal specifications* *Extract on dry carrier (%)*	*Vol. oil in extract vol/wt. (%)*
1	*2*	*3*	*4*	*5*	*6*	*7*
Ginger	S.R.	1.134	4.000	28	4.0	25
Mace	E.O.	1.418	5.000	100	–	–
	O.R.	1.985	7.000	40	–	–
	S.R.	1.985	7.000	50	7.5	50
Marjoram, sweet	E.O.	0.140	0.494	100	–	–
(special)	O.R.	2.268	8.000	4	–	–
Marjoram, sweet	S.R.	0.709	2.500	40	3.0	40
Mustard	O.R.	1.276	4.500	5	–	–

Contd . . .

Table 5.3 Contd . . .

Extractive	*Type*[a]	*g/Type equivalent to 1 oz ground spice*	*Extractives, on soluble, dry, edible carrier (%)*	*Minimum vol. oil in extractive vol/wt. (%)*	*Federal specifications* Extract on dry carrier (%)	*Federal specifications* Vol. oil in extract vol/wt. (%)
1	2	3	4	5	6	7
Nutmeg	E.O.	1.134	4.000	100	–	–
(special)	O.R.	2.552	9.000	60	–	–
Nutmeg	S.R.	1.701	6.000	80	6.0	80
Onion oil	0.012	0.0425	100.00	-	–	
Onion (special)	O.R.	0.071	0.250[d]	5g	–	–
	O.R.	0.284	1.000[e]	5g	–	–
Oregano	E.O.	0.212	0.748	100	–	–
(special)	O.R.	1.134	4.000	17	–	–
	O.R.	0.851	3.000	27	–	–
Paprika (2500 color units)	O.R.	2.268	8.000	0	–	–
(1,000 color units)	O.R.	2.268	8.000	0	–	–

Contd . . .

Table 5.3 Contd . . .

Extractive	Type[b]	g/Type equivalent to 1 oz ground spice	Extractives, on soluble, dry, edible carrier (%)	Minimum vol. oil in extractive vol/wt. (%)	Federal specifications: Extract on dry carrier (%)	Federal specifications: Vol. oil in extract vol/wt. (%)
1	2	3	4	5[c]	6	7
Parsley (special)	O.R.	0.094	0.333[d]	12	–	–
	O.R.	0.851	3.000[e]	12	–	–
Pepper, black, Indian	O.R.	1.488	5.250	23[f]	4.5	150
Indian decolorized	O.R.	1.488	5.250	23[f]	–	–
Rosemary	E.O.	0.142	0.500	100	–	–
(special)	O.R.	1.559	5.500	10	–	–
Sage, Dalmatian	E.O.	0.350	1.235	100	–	–
(special)	O.R.	2.126	7.500	25	–	–
Sage, Dalmatian	S.R.	0.710	2.500	65	5.4	65

Contd . . .

Table 5.3 Contd . . .

Extractive	Type[b]	g/Type equivalent to 1 oz ground spice	Extractives, on soluble, dry, edible carrier (%)	Minimum vol. oil in extractive vol/wt. (%)	Federal specifications: Extract on dry carrier (%)	Federal specifications: Vol. oil in extract vol/wt. (%)
1	2	3	4	5	6	7
Tarragon (estragon)	E.O.	0.035	0.123	100	–	–
	O.R.	0.709	2.500	12	–	–
Thyme	S.R.	1.134	4.000	50	4.0	50
Turmeric	O.R.	4.394	17.500	0	–	–
standard 7% curcumin						
super 14% curcumin	O.R.	2.410	8.500	0	–	–

Table 5.3 contd. . .

Notes:—

[a] Spice equivalencies are based on oils and oleoresins produced and supplied by Fritzsche, Dodge & Olcott, Inc., New York. Materials from other suppliers may be similar or differ considerably in their values.

[b] E.O. = Essential oil. O.R. = Oleoresin. S.R. = Superesin.

[c] Special = A selective diluent has been added (propylene glycol or a fatty oil) to ease the dispersibility of the oleoresin. Onion and garlic oleoresins come in water-or oil-dispersible forms.

[d] = Fresh spice equivalent.

[e] = Dehydrated spice equivalent.

[f] = 53% piperine content minimum.

A specific name has to be used for ingredients in the list of ingredients except that for ingredients falling in the respective classes the class titles may be used, namely edible fats, edible oil, spices and condiments, edible starches (except modified starches), vitamins and minerals, salts. However in the case of curry powder and mixed masalas, whole or other such masalas containing spices, either whole or powdered, as major ingredients, the names of spices used in the product have to be mentioned on the package in descending order of their composition by weight and so on.

Unfortunately some manufacturers mislead their competitors deliberately by not following the above labelling regulations and some of the individual ingredients in order to alter the aforesaid descending order of ingredients on the label. The public analyst and the industrial analyst is expected to decode the correction and accuracy of the label declaration, the former for the purpose of enforcement of food laws, the latter to find out the merits of the competitors product. This is achieved by proceeding according to the following guidelines intelligently and not mere mechanically.

The product is examined critically for uniformity, texture, colour, the type of labelled ingredients, the method of using the product, etc.

Crystalline If the product is crystalline, uniformly coloured, no organic extraneous matter like chopped stalks of plant materials, it can be inferred that extractives and not the parent materials have been employed. Uniform large particles suggests use of coarse corn syrup solids as a base. Small to fine crystalline ingredients means the presence of salt/dextrose/fine corn syrup solids / hydrolysed cereal products.

Taste If the product tastes sweet, it may be due to dextrose. Sweetness without the cooling effect means absence of dextrose and presence of sugar or corn syrup solids is indicated particularly if the sweetness level is very low. A slight cooling effect with pronounced sweetness may be due to presence of dextrose and sugar. Saltiness can also be measured. Absence of salt, dextrose, sugar but persisting low sweetness may be due to hydrolysed cereal solids.

Shiny, needle like crystal are typical of mono sodium glutamate (MSG). Many substances interfere, hence its presence defies detection merely through taste alone unless it is one of the predominant ingredients. Its excessive use may reduce the overall flavour. Its optimum use intensifies and enhances the natural flavours, checkmating their fading, modulates onion's pungent notes and bitterness of bitter gourd, etc. MSG produces a pleasant mouthfeel, an overall rich flavour, reduces earthiness making the product relishable.

Ribonucleotides $(Na)_2$ 5 inosinate as well as gunylate are similar to MSG even at 10% of MSG level. Together they are synergistic. Like MSG, a sense of viscosity is apparent through a better mouthfeel along with a sensation of meatiness.

Hot and pungent taste is the result of the presence of red/black/white pepper or ginger. White pepper affects the back portion of the mouth, black pepper the front portion and red pepper will affect the entire posterior oral cavity and the lips. Black pepper extract alone makes the seasoning appear greenish-brown.

Similarly ginger extract shall impart a brownish or a capsicum-red shade. In a complex mixed seasoning, the colour evaluation does not help much.

Oleoresins of capsicum and paprika get oxidised in the presence of salt within a short time, necessitating the use of antioxidants. These extractives in contact with aluminium in salt's presence impart grayish-blue colour to seasoning rendering it unfit for further use. This reaction can take place within a week's time. Reddish colouration suggests the presence of capsicum, paprika or beet powder. Yellowish or brownish yellow suggests the presence of turmeric, annatto, mustard or even saffron, although saffron is rare.

Tastewise, sugar and salt offset each other if in equal proportions. This technique increases unit weight of the seasoning for a given weight of meat increasing the bulk cheaply.

Mustard is used to simulate natural spices in a seasoning especially when only extractives have been employed.

Geographical origin of the spice blend/seasoning has its peculiarity, the analyst should look for regional clues in the formulations.

Screen analysis helps in estimating fairly closely the proportion of various ingredients present in spice blends containing mixture of spices, base/other materials. Onion, pepper, chilli, coriander, mustard, MSG particles are easily discernible. For confirmation, tasting the individual particles helps. Accurate weighing of the test material and various sieved fractions gives another important clue to the final composition.

Methodology of identifying the ingredients through flavour/ aroma by dissolving the test material in hot water (98°C) then cooling to 74°C, then to 30°C, as explained in the beginning is also very helpful in decoding the formulations.

On the basis of above data, declaration on the label can be fairly scrutinized. If duplication of a competitor's product is desired that also can be done. Under Rule 63, there is no specific names/groups/ which have been permitted for use as flavours and/or flavouring agents, the question of their limits does not arise even. In other words their quantities to be used in food processing is limited by their ability to impart and enhance the flavour of the final processed food product.

Nevertheless, menthol, vanilla and monosodium glutamate (MSG) are of interest as these are extensively used in various food products. Menthol is used mainly to flavour confectionery and *pan masala.*

Vanillin is extensively used in ice creams and monosodium glutamate to enhance flavour of meat preparations, soups, etc.

Under Rule 63A, the use of following flavouring agents is prohibited in any article of food,

(i) Coumarin and dihydrocoumarin
(ii) Tonkabean (Dipteryl odorate).
(iii) p-asarone.
(iv) Cinnamyl anthracilate.
(v) Estragole
(vi) Ethyl methyl ketone
(vii) Ethyl-3-phenylglycidate
(viii) Eugenyl methyl ether

(ix) Methyl β napthyl ketone

(x) P. propylanisole

(xi) Saffrole and isosaffrole

(xii) Thujone, isothujone, α and β thujone.

Their profiles are as follows:

(i) ***Coumarin*** Its synonyms are cis-o-coumarinic acid lactone, cumarin, coumarin anhydride and tonkabean camphor. It is present in lavender oil, tonka beans, wood *(Asperula species)*, in sweet clover (melilotus). It occurs as orthorhombic, rectangular plates.

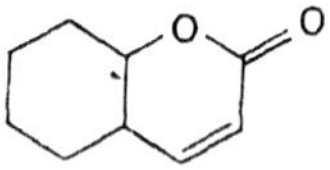

Figure 5.2

It has pleasant, fragrant odour resembling that of vanilla beans, burning taste, mt. pt 69 ± 1°C, boiling pt. 289 ± 1°C. One gm dissolves in 400 ml cold, 50 ml. boiling water; freely soluble in alcohol, chloroform, ether, oils and also in alkali hydroxide solutions. LD_{50} orally in rats, guinea pigs-680, 202 ppm and *is a permitted pharmaceutic aid for flavouring yet is banned for use in food articles under Rule 63-A.*

(ii) ***Cinnamyl Anthranilate*** It is also described as 2-amino benzoic acid 3-phenyl-2-propenyl ester; 3-phenyl-3-propenyl anthranilate, cinnamyl O-amino benzoate.

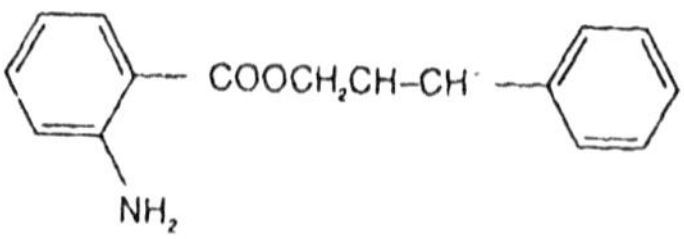

$COOCH_2CH-CH$

NH_2

Figure 5.3

It is a synthetic imitation of grape or cherry flavour, crystals having mt. pt. 61-61.5°C. Carcinogenic studies of this flavour were reported by Stoner *et.al. Cancer Research,* 1973,33,3069, hence has been banned.

(iii) ***β Asarone*** Asarones also known as asarin; asarum camphor, or as a rabacca camphor is obtained from the roots of *Asarum*

europaeum (L); *Arista locliaceae* by distillation with water and in *Acorus calamus (L).* It occurs in nature as a mixture of two isomeric forms, α is the-trans-and β is the-cis-form. The unqualified term asarone is often used for α-asarone.

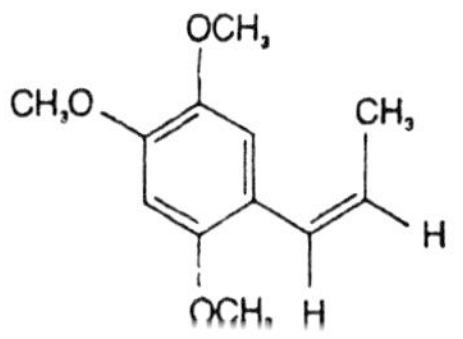

OCH_3 CH_3O CH_3 H OCH_3 H

β–asarone

Figure 5.4

Its insect chemosterilant effects have been studied by B.P. Saxena, *et. al.; Nature,* 1977,270, 512 hence has been banned.

(iv) ***Tonkabean*** It is the fruit of the shrubby leguminous tree, *Dipteryodorata* containing a single seed with a pleasant odour used in perfumes and *as a vanilla substitute,* and in flavouring tobacco.

(v) ***Estragole*** Synonyms are p-allylanisole; chavicol methyl ether and esdragol.

Estragole is the main constituent of tarragon oil, also known as estragon oil, the oil from *Artemisia dracunculus where* it occurs to an extent of 60-75%. It also occurs in pine oil and in American turpentime oil.

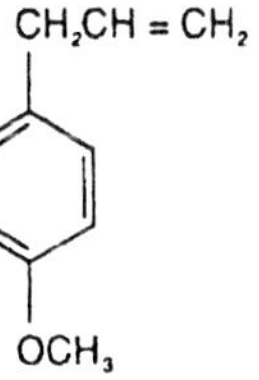

$CH_2CH = CH_2$

OCH_3

Figure 5.5

It is a liquid, $d\frac{2}{4}$ 10.9645, its boiling points vary with the determination pressures viz. at 764 it is 216°C, at 25 it is 108-114°C, at 12 it is 95-96°C, refractive index $n_D^{17.5}$ 1.5230. It is soluble in alcohol, chloroform, forms azeotropic mixtures with water. It is used in perfumes and as flavour in foods and liquors.

(vi) ***Ethyl Methyl Ketone*** Synonyms are 2-butanone; methyl ethyl ketone; MEK and 2-oxo butane. $CH_3CO\ CH_2\ CH_3$. It is a flammable liquid having acetone like odour. Density d_4^{20} is 0.805, melting point 86°C, boiling point 79.6°C. Flash point (cup. method) (-6°C). n_d^{15} 1.3814. It is soluble in about 4 parts water (27.5%), less soluble at higher temperature, miscible with alcohol, ether, benzene. It forms a constant boiling mixture with water, b.pt. 73.4°C, contains 88.7% methyl ethyl ketone. Solubility of water in methyl ethyl ketone is 12.5% at 25°C.

(vii) ***(a) Safrole*** It is also know as allylcatechol methylene ether; allyl dioxybenzene methylene ether, m-allyl pyrocatechin methylene ether. It is a constituent of several essential oils, notably of sassafras in which it is present to the extent of about 75%. It is a colourless or slightly yellow liquid, sassafras odour, d_4^{20} 1.096, mt. pt. about 11°C and b.pt 232-234°C. d_D^2 1.5383. Insoluble in water and very soluble in ether. Miscible with chloroform and ether.

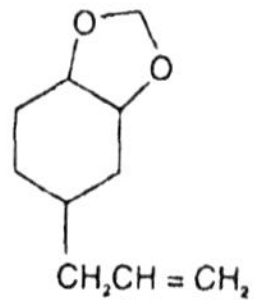

$CH_2CH = CH_2$

Figure 5.6

Safrole is a constituent of *Cinnamomum cecidodaphne* (Meison), *C. zeylanicum (Breyn)* leafoil or cirnnamon leafoil, oil of nutmeg and mace (0.6%) *Myristica fragrans* (Hout) and camphor oil

(b) ***Isosafrole*** Trans form is liquid and has odour of anise. Its boiling point varies with pressure, at 760 mm of (Hg) 253°C, at 100 mm of (Hg) 179.5°C, at 20 mm of (Hg) 135.6°C, at 3.4 mm of (Hg) 85-86°C, mt-pt 8.2°C, d_4^{20} 1.1182, n_D^2 1.5691, UV_{max} (96% alcohol) 305, 267, 259.5 nm. Miscible with ether, alcohol and benzene. It is soluble in 8 parts of 90% alcohol.

Cis form is also liquid, b. pt at 3.5 mm of (Hg) 77-79°C; mt pt -21.5°C, d_4^{20} 1.1182, n^2_D 1.5691, UV_{max} (96% alcohol) 296.5, 259 nm.

In small quantities together with methyl salicylate in root bear or *sarsaparilla flavours.*

(viii) Thujone It is also referred as 3-thujanone. It is a constituent of many essential oils; present in thuja, etc. *(Thuja orientalis, morpankhi).* Equilibrium mixture contains 33% α-thujone and 67% p-thujone. α-and β-thujones differ only in the stereochemistry of the 4-methyl group. Thujone is colourless or almost colourless liquid. UV_{max} (iso octane) 300 nm (e23).

Practically insoluble in water, soluble in alcohol and many other organic solvents. Ingestion may cause convulsions. L-form, a-thujonc; boiling point 83.8.

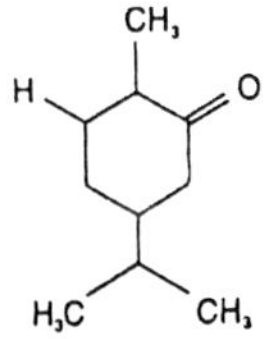

Figure 5.7

84.1°C, d_4^{20} 50.9109, n_D^{15} 1.4490, α_D^{20} - 19.2°,D-form,d-isothujone, β–thujone, boiling point 85.7-86.2°C; d_4^{25} 0.9135, n_D^{25} 1.4500,

(ix) **Methyl β-napthyl ether:–** Also known as 2-methoxy napthalene, nerolin old; yarayara. $C_{11}H_{10}O$, mol wt 158.19.

Leaflets from ether, mt pt 72°C, b pt 272°C. Practically insoluble in water, sparingly soluble in alcohol, soluble in ether CS_2, benzene.

Under Rule 64, diethlene glycol or mono ethyl ether shall not be used as a solvent in flavours.

This necessitates their detection but estimation is done only for menthol as a good manufacturing process and for MSG under statutory obligation of Rule 64-B, according to which the total glutamate content of the ready-to-serve food does not exceed 1%. It shall not be added to any food for use by the infants below twelve months.

GC determination of menthol sweets and pan masala Menthol along with other flavouring matter is steam distilled into $CHCl_3$. The dry $CHCl_3$ extract is directly subjected to GC on 10% carbowax-

20 M using FID. The amount of menthol present in the sample is determined using peak height/area of sample and the standard.

Apparatus

(i) CIC Model GC filled with FID or equivalent.

(ii) Stainless steel column (183 cms × 0.3175 cm O.D.) packed with 10% carbowax, 20 M on Chromosorb - W (90 ± 10 mesh)

Conditions

(i) Column temperature 185°C.

(ii) Detector and injector temperatures 240°C

(iii) Nitrogen as carrier gas @ 25 ml/minute.

Under these conditions retention time (R_T) for menthol is 3.0 minutes.

Reagents

(i) *Standard solution* (0.5 mg/ml) menthol solution in $CHCl_3$.

(ii) $CHCl_3$ of chromatographic grade.

Procedure

To about 10 g of sample (accurately weighed) in a distillation flask is added 20 ml H_2O. The flask is connected to a steam generator and condenser. The other end of the condenser is dipped in a conical flask containing 18 ± 2 ml of $CHCl_3$. The contents are steam distilled slowly until about 50 ml distillate is collected into conical flask. The distillate is transferred to a 100 ml separating funnel, $CHCl_3$ layer is separated. The aqueous phase is extracted with 2 x 10 ml portions of $CHCl_3$(ii) above. The combind $CHCl_3$ extracts are dried over anhydrous Na_2SO_4, the volume made up to 50 ml. If the menthol concentration is less in the extract, it is concentrated under nitrogen to 10 ml.

1 µl of standard menthol solution in CH Cl_3 is injected into GC and peak height/area is recorded. Similarly 1 µl of sample solution in $CHCl_3$ is injected, R_f, being same as of standard its peak height / area is also recorded. This procedure is done in duplicate. The menthol content of the sample is determined from average peak height/area of standard versus sample.

TLC Detection of Vanillins and Coumarins

Vanillin, ethyl vanillin, coumarin and dihydro coumarin in vanillin extracts are separated on TLC and detected by spraying hydrazine sulphate HCl, KOH, diazotised sulfanilic acid respectively.

Reagents

(i) Developing Solvent System

  (a) Pet. ether : ethyl acetate :: 2:1

  (b) Hexane : ethyl acetate :: 5:2

(ii) Spray Reagents

  (a) 1% hydrazine sulphate in IN HCl

  (b) Cone, methanolic solution of KOH

  (c) 1% diazotised sulfanilic acid.

(iii) Silica gel G.

Procedure

The sample extract in ethyl acetate is spotted on TLC plate coated with (iii). The TLC developing chamber is saturated with either of the (i), the plate is developed to about 12 cms height, air dried and sprayed with (iia). Another plate is developed and sprayed with (iib) and viewed under uv and day light. The previously sprayed plate with (iib) is again sprayed with (iic) and viewed under day light.

Presence of above compounds is confirmed from the Rf of standards. For general information Table 5.4 is a good ready reference.

Table 5.4: *TLC of Silica gel G*

S. No.	*Compound*	*Rf Value*		*Detection*				*KOH and thereafter diazotised sulfanilic acid*
		Solvent A	*Solvent B*	*Hydrazine sulphate day light*	*uv*	*KOH day light*	*uv*	
1.	Vanillin	0.30	0.27	Yellow	Orange	-	-	Rose
2.	Ethyl Vanillin	0.46	0.42	Yellow	Yellow	-	-	
3.	Coumarin	0.50	0.38	-			Green	Orange
4.	Dihydro Coumarin	0.62	0.50	-				Orange

Determination of Caffeine, Benzoate and Saccharin in soda beverages Saccharin, benzoate and caffeine are simultaneously quantified in soda beverages by liquid chromatography on U-Bondapak-C_{18} column using acetic acid (20%) buffered to pH 3.0 with saturated Na acetate and modified by adding 0.2% isopropanol. The concentration of the additives in the sample is determined by measuring the peak heights using an UV absorbance detector monitored at 254 nm.

Special Apparatus

(i) Water Associates liquid chromatograph equipped with 6000A solvent delivery system and model U6 K injector or equivalent.

(ii) 10 mv strip chart recorder (Houston Omni Scribe or equivalent)

(iii) U-Bondapack C_{18} column, 300 x 4 (i.d.) mm, flow rate 2 ml/minute.

(iv) UV 254 detector, its sensitivity adjustable from 0.02-0.05 AUFS.

Reagents

(a) Standard solutions Individual standard solutions are prepared from standard compounds to get the following concentrations.

(i) Sodium saccharin, 0.5 mg/ml

(ii) Caffeine, 0.5 mg/ml

(iii) Sodium benzoate, 0.5 mg/ml.

These solutions are used to determine sensitivity for detector response and retention times of individual standards as well as to optimise LC conditions for complete resolution and to quantify by mixing all these into a mixed solution.

(b) Mobile Phase 20% acetic acid (v/v) is buffered to pH 3.0 with saturated Na acetate solution modified with 0.2% isopropanol to obtain base line resolution and retention times of standards from standard solution (a) above, in approximately 10 minutes. It is degassed prior to use.

Preparation of Sample

(i) Carbonated beverages are decarbonated by agitation or ultrasonic treatment. If free of particulate matter, it is injected directly.

(ii) Beverages containing particulate matter are filtered through milipore filter (0.45 μm). First few ml are discarded. If large amount of particulate matter is present the sample is centrifuged prior to filtration. Thus filtrate is directly injected.

Determination Standard solutions a (i), (ii) and (iii) are mixed. 10 μl or about but known volume is injected in duplicate, subject to peak heights agreeing within ± 2.5%. A known volume of the prepared sample (i) or (ii) above, is injected in duplicate. Their peak heights are measured:

Percentage of compound = $C_1 \times H/H_1 \times V/V \times 0.1$, where, C_1 = Concentration of standard in mg/ml.

H and H_1 = average peak heights of sample and standard respectively.

V and V_1 = volume injected in μl of sample and standard respectively.

Detection of Safroles This method is quantitative for CH_3-salicylate and of semiquantitative value for other compounds. All solvents must be chromatographically pure and confirm to the following test.

500 gm of solvent is concentrated to 10 ml in an evaporative concentrator and chromatographed on the prepared column with instrument set at maximum sensitivity to be used during analysis. 5 ul of this concentrated solvent must not show any trace of peaks beyond solvent front. Distilled-in-glass solvents generally conform to test standards. Those solvents not passing this test must be purified as described ahead.

1. ***Reagents***

(a) ***Methanol (anhydrous)*** 1 litre of methanol is refluxed with 10 gm of KOH and 25 gm zinc dust for 3 hours then distilled. discarding the first 100 ml of the distillate.

(b) ***Chloroform*** It is redistilled at its boiling point, the distillate is stored under 1% methanol.

(c) ***Flavour Standard Solution***

(i) Stock Solution 79 mg/ml (1%), 7.90 gms of the flavour compound is diluted to 100 ml with methanol (a),

(ii) Working Standard 7.9 mg/ml. It is prepared by diluting 1 ml of (c) (i) to 10 ml with (a).

(d) ***Internal standard solution*** n-decylalcohol is used for the determination of CH_3-salicylate and m-tolyl acetate for the determination of other compounds. Stock and working standard solutions are as at (c) (i) and (ii) above.

(e) ***Defoaming agent*** Dow Corning antifoam A or equivalent.

2. Apparatus

(i) ***Steam distillation apparatus*** It is a RB flask of 500 ml capacity with a long neck and 24/40 tapering joint. There is an adapter with a 24/40 tapering joint at bottom and at side at 75° angle and with 24/40 joint at top; a gas inlet tube, 30 cms., 24/40 inner joint and perforated 8 mm bulb at dispersion tip; condenser, 200 mm long with 24/40 joint and adapter; vacuum take off type with extended lower tube, straight 24/40 joint.

(ii) ***Concentrator*** It is an evaporative, Kudema-Danish type, 500 ml, with 10 ml receiving flask.

(iii) ***Gas chromatograph*** It has flame ionization detector, and operating parameters as,

Glass column, 1.8 m x 4 mm (i.d.) coiled or u-shaped, packed with 15% polypropylene glycol adipate (Reoplex 400) on 60-80 mesh Gas Chrom P (w/w); temp (°) _ column (isothermal) detector 220 injection port 180; nitrogen carrier gas 40 *psi* @ 90ml/minute. Column is programmed over range 90-150° @ 2°/minute.

3. Determination of relative retention Time (RRT)

(a) ***CH_3 -Salicylate*** 1 ml each of (c) ii) and (d) n-decyl alcohol is pipetted into 10 ml volumetric flask. The volume is made up with methanol 1(a) above.

(b) ***Other flavour compounds*** 1 ml of appropriate working standard solution is pipetted along with 5 ml of m-tolylacetate working internal standard solution into 10 ml volumetric flask. The volume is made up with methanol 1 (a) above.

4 ± 1 ul is injected into gas chromatograph. Sensitivity and attenuation controls are set to provide peak height not more than 80% of chart scale.

RRT is time for standard component peak to emerge divided by the time for internal standard peak to emerge. Time is measured from emergence of solvent front. RRT is used to identify component peak in sample. Peak Area (PA) is calculated.

4. Determination

Samples must be analysed on the same day and under the same conditions used for determination of RTT (3) above.

(a) ***CH_3 Salicylate*** 200 ml of sample, degassed if necessary, is placed in 500 ml R.B. flask. 10 ml of methanol 1 (a), few SiC chips and small amount of defoamer 1 (e) is added, followed by 1 ml n-decyl alcohol working standard 1 (d). The flask is attached to 2 (i) and not connected to steam generator. The sample is brought to incipient boiling, steam generator is now connected and @ 2 ml/ minute. 90 ml of distillate is collected in a 100 ml Nessler tube or graduated tube containing 10 ml methanol to cover delivery tube outlet. The condenser and delivery tube are rinsed with 5 ml of methanol. The distillate is quantitatively transferred to a 250 ml separating funnel (SF), 50 ml of saturated NaCl solution is added, contents mixed, then extracted with 25, 25 and 10 ml of 1 (b) by carefully shaking for 5 minutes for each extraction. 1(b) is filtered through 12.5 cms Whatman No. 30 filter paper, containing 4 ± 1 g Na_2SO_4 into a 500 ml 2(ii). Both SF and paper are washed with 10-15 ml 1(b) into 2(ii), a few SiC chips are added and evaporated on steam bath to nearly 7 ml. After cooling, the contents are diluted to 10.0 ml with l(b). CH_3-salicylate is estimated as:

Calculation

mg/ml or (ppm) CH_3-salicylate=(PA/PÁ) (PA_1/$PÁ_1$) C × F, where PA and $P\bar{A}$ are peak areas of sample and standard solutions respectively; $PÁ_1$ and PA_1 are peak areas of internal standard in standard and sample solutions respectively. C = concentration of CH_3-salicylate in working standard solution (C ii) and F = dilution factor (1/20)

(b) ***Other flavour compounds*** The same procedure as at 4(a) except that 5.0 ml m-tolylacetate working internal standard 1 (d) is used and 150 ml distillate @ 5ml/minute is collected.

Detection of Tkujones To 500 ml alcohol sample 1 ml of freshly distilled aniline is added along with 1 ml $H_3 PO_4$, the contents are refluxed on a steam bath for 30 minutes and then distilled. First 100 ml distillate is rejected. To the next 100 ml distillate 0.5 g semicarbazide hydrochlorides 600, mg of anhydrous sodium acetate and 1.0 gm crystallized salt is added, the mixture is then left aside for overnight. Alcohol is distilled off at minimum pressure. The residue is steam-distilled to remove essential oils and other volatile materials, the distillate is collected. First 15 ml of this distillate is rejected. The condenser is washed down with little alcohol then with water. The sample is cooled, 1 ml of (1:1) H_2SO_4 is added and again steam distilled, collecting 20 ml distillate in cylinder then poured into a small separating funnel. 20 ml ether is measured in the same cylinder, this too is poured in the separating funnel. Contents are shaken and the ether solution is separated. To it is added 10 ml of 65% alcohol, then ether is let to evaporate spontaneously. After all ether evaporation, the odour of the residue is noted. Odour of thujone will be apparent if more than 2 mg is present in the solution, provided it is not masked by the presence of other odouriferous substances. To perform the modified legal test, to the above obtained solution 1 ml of 10% $ZnSO_4$ solution and 0.25 ml of freshly prepared aqueous sodium nitroprusside solution (0.1 g/ ml) is added. Slowly and with constant stirring 2 ml of 5% NaOH solution is added, the reactants are let to stand for about two minutes. 1.5 ml of acetic acid is mixed. A precipitate of raspberry red colour, resembling alcohol precipitate of red fruit juice, shows presence of thujone. Negative test is shown by similar precipitate having appearance similar to that of alcohol precipitate from apple jelly or other light coloured fruit.

This test has been adopted by *AOAC(1980) from JAOAC,* 1936, 19, 120 and 1937, 20, 69.

MSG-A FLAVOUR ENHANCER

A flavour enhancer is a substance that is added to a food to supplement or enhance its original taste or flavour. The term *flavour potentiator* has also been used with the same meaning. The most commonly used substances in this category are monosodium L-glutamate (MSG), disodium, 5-inosinate (IMP), and disodium, 5-guanylate (GMP).

It was Prof. Kikunao Ikeda (1908) who isolated the essence of "tastiness" of soup, i.e. glutamic acid from *kombu (Laminaria japonica)* bouillon, and suggested that this should be a basic taste independent of the four traditional basic tastes: sweet, sour, bitter and salty. Glutamic acid itself was first isolated from gluten (wheat protein) by Rithausen (1866). Today about 400,000 tons of MSG are manufactured annually in some 15 countries throughout the world.

MSG alone did not yield the complete taste of soups and the search continued and culminated in 1960 when two other ingredients, inosinic acid and, 5-guanylate, were discovered as the missing links. Contextually these both and glutamates are key components of living organism.

Glutamate is naturally present in virtually all foods, including meat, fish, poultry, milk and many vegetables. Protein rich foods such as human milk, cow's milk, cheese, and meat contain large amounts of bound glutamate. Whilst most vegetables contain little, yet mushrooms, tomatoes and peas have high levels of free glutamate. Glutamate is an important element in the natural and traditional ripening processes to achieve fullness of taste in foods hence the wide spread use of glutamate (natural as well as of synthetic) in popular cuisines of the world.

Nucleotides are specifically distributed. 1 MP is dominant in meat, poultry and fish, whereas adenosine monophosphate (AMP) is dominant in most crustaceans and molluses further more, almost all vegetables contain AMP. The GMP content of mushrooms is particularly high.

Psychometric studies on aqueous solutions of the four basic tastes revealed that except sweetness, the rest three are rated as unpalatable over a wide concentration range. Similarly, MSG in aqueous solution had an unpleasant rating or was rated neutral in acceptability at all concentrations. Further there is an optimal concentration for MSG added to food. Beyond this most palatable concentration, the palatability of food decreases. Thus the use of MSG is self limiting as its overuse decreases palatability.

The major use of MSG in cooking around the world is as a flavour enhancer in soups and broths, sauces and gravies, flavourings and spice blends. MSG is also included in a wide variety of canned and frozen meats, poultry, vegetables and combination dishes.

Results of taste panel studies indicate that a level of 0.05 to 0.8% by weight in food gives the best enhancement of the food's natural flavour. Because MSG is readily soluble in water, recipes often call for dissolving in the aqueous ingredients of the products such as salad dressings before they are added to food.

Toxicology of Glutamate The acute toxicity of glutamate is very low.

Glutamate in pediatric neurology Glutamate is one of the predominant aminoacids of the mature brain. The high concentration and extensive metabolism of glutamate as well as its derivatives are among the most significant aspects of brain metabolism.

Glutamate links aminoacid metabolism with the tricarboxylic acid (TCA) cycle. It is a vital component in the synthesis of proteins and peptides. It is involved in the detoxification of ammonia and recent evidence points to its role as an energy supply during hypoglycemia.

Recent research in neuroscience has got increasing evidence that glutamate and γ-amino butyric acid (GABA) are important neurotransmitters. Glutamate receptors in the brain are associated with trans membrane ion channels which specifically open when glutamate and/or other specific molecules are bound. Mechanism of normal glutamate mediated nerve signal transduction involves the influx of ions like Na^+ and Ca^{++} into the neurons and the resulting depolarization of the membrane potential.

During normal functioning, glutamate released as a neurotransmitter into the synaptic space is efficiently removed by cellular uptake mechanisms which can handle very large amounts of glutamate. Rapid removal is evidenced by the extremely large doses of glutamate required to produce even a minimal neutrotoxic reaction even when it is injected directly into the brain.

When the uptake mechanism is overwhelmed by intracranial administration of massive doses of glutamate, neuronal injuries can be induced. This toxicity is thought to be mediated by glutamate receptors but the mechanism is not well established for all types of receptor classes.

Unlike the essential, neutral or basic amino acids, the acidic amino acid, glutamate, is synthesized in the brain cells at rates commensurate with metabolic demands. Glutamate levels in the brain are higher than plasma glutamate levels.

The blood-brain barrier for glutamate is somewhat different from other aminoacids. It has been reported that afflux of glutamate from the brain is about 7 times greater than its influx to compensate for the large production of its amnio acid in the brain. Adding MSG to food does not sufficiently elevate plasma glutamate levels in infant and adult humans to anticipate any significant increase in brain glutamate levels.

Historically, substantial amounts of glutamate have been administered with the intention of treating subjects with various medical conditions. No major complaints due to glutamate ingestion have been reported.

MSG has been used in the treatment of mentally retarded children. High doses of MSG (10-15 gm) had no effect on BMR, EEC, ECG, BP, respiration and heart rate over a period of eleven months. Studies on epileptic children have been non-conclusive.

On the other side, number of care reports and few studies are available, which describe various neurologic symptoms related to MSG tolerance. Children seem to describe similar symptoms of CRS (Chinese restaurant syndrome), like burning sensation, facial pressure and chest pain with the same degree of prevalence. Other adverse reactions reported to MSG, include shuddering attacks, migrane-like syndrome and psychiatric symptoms.

Anecdotal reports of acute psychiatric reactions like intermittent hyperactivity, aggression and longer term psychologic effects which lasted 1-2 weeks, like depressed, motor slowing, doubt ridden and gloomy fantasies, have been reported.

Large single dose of MSG results in its high plasma levels hence caution.

Perhaps these factors led to the stipulation. Under Rule 42(s) that "MSG shall not be added to any food for use by infant below twelve mondis".

This is despite the fact that studies in the developed countries had indicated that the expected daily intake of MSG on person-days vary from 0.3 mg/kg in 0-5 months age to as high as 6.8 mg/ kg for 12-23 months age child[12].

Use of Monosodium Glutamate.—Under Rule 64 B of PFA Rules, 1955, as amended w.e.f. 21.9.2005 monosodium glutamate may be added to foods as per the provisions contained in Appendix C, subject to Good manufacturing Practice (GMP) level and under proper label declaration as provided in rule 42(S). It shall not be added to any food for use by infant below twelve months and in the following foods:—

(List of foods where Monosodium Glutamate is not allowed).

1. Milk and Milk Products including Buttermilk.
2. Fermented and renneted milk products (plain) excluding dairy based drink.
3. Pasteurized cream.
4. Sterilised, UHT, whipping or whipped and reduced fat creams.
5. Fats and Oils, Foodgrains, Pulses, Oil seeds and grounded/ powdered foodgrains.
6. Butter and concentrated butter.
7. Fresh fruit.
8. Surface treated fruit.
9. Peeled or cut fruit.
10. Fresh vegetables.
11. Frozen vegetables.
12. Whole, broken or flaked grains including rice.
13. Flours of cereals, pulses and starches.
14. Pastas and noodles (only dried products).
15. Fresh meat, poultry and game, whole pieces or cuts or comminuted.

16. Fresh fish and fish products, including mollusks, crustaceans and echinoderms.
17. Processed fish and fish products, including mollusks, crustaceans and echinoderms.
18. Fresh eggs, Liquid egg products, Frozen egg products.
19. White and semi-white sugar (sucrose and saccharose, fructose, glucose (dextrose), xylose, sugar solutions and syrups, also (partially) inverted sugars, including molasses, treacle and sugar toppings.
20. Other sugars and syrups (*e.g.* brown sugar and maple syrup).
21. Honey.
22. Salt.
23. Herbs, spices and condiments, seasoning (including salt substitutes) except seasoning for Noodles and Pastas, meat tenderizers, onion salt, garlic salt, oriental seasoning mix, topping to sprinkle on rice, fermented soybean paste, Yeast.
24. Infant food and Infant milk substitute including infant formulae and follow-on formulate.
25. Foods for young children (weaning foods).
26. Natural Minerals water and packaged drinking water.
27. Concentrates (liquid and solid) for fruit juices.
28. Canned or bottled (pasteurized) fruit nectar.
29. Coffee and coffee substitutes, tea, herbal infusions, and other cereal beverages excluding cocoa.
30. Wines.
31. Margarine.
32. Fat Spread.
33. Fruits and Vegetables products except those where Monosodium Glutamate is permitted under Appendix C of these rules.]
34. Carbonated Water.
35. Baking Power
36. Arrowroot.
37. Sago.
38. Plantation Sugar, Jaggery and Bura.
39. Ice-Candies.

40. Ice creams and Frozen deserts.
41. Cocoa Butter.
42. Saccharine.
43. Malted Milk Food and Milk based foods.
44. Bread.
45. Vinegar.
46. Sugar Confectionery, Toffee, Lozenges.
47. Chocolate.
48. Pan Masala.
49. Alcoholic Beverages.

Determination of Monosodium Glutamate (MSG) in foods
Glutamic acid is extracted from foods using water, separated from other amino acids by ion-exchange chromatography and titrated potentiometrically using 0.1 NaOH. The chromatographic column should have 55 x 22 mm (o.d.) tube, 30 ml bed volume with Dowex 50WX 8(H form) 100-200 mesh, the activated carbon may be of Darco -G. 660 or equivalent.

Preparation of Samples

(a) ***Products in dry form*** 40 g sample is reduced to powder in mortar. 10 g of this powdered sample is weighed in a 250 ml beaker.

(b) ***Undiluted concentrated soups or canned green beans*** The entire sample is homogenized in a blender, out of which 20 g are taken in a 250 ml beaker for analysis.

(c) ***Consomme type (clear condensed) soup*** 20 g is weighed into a 250 ml beaker.

Nevertheless, the above prepared sample is diluted to about 70 ml with water at room temperature, mixed until ali water soluble substances are in solution. 6 g of above activated carbon added and mixed thoroughly, for products containing starch 60 ml acetone is also added to precipitate starch which assists in making a solution of the sample. After 30 minutes the contents are filtered under vacuum through 60 ml coarse fretted glass funnel containing asbestos pad. The flask is washed with 6 × 25 ml portions of acetone : H_2O :: 1:1. The filtrate and washings are collected in 400 ml beaker, 2 drops of HCl (1 + 25), contents are evaporated on a steam bath

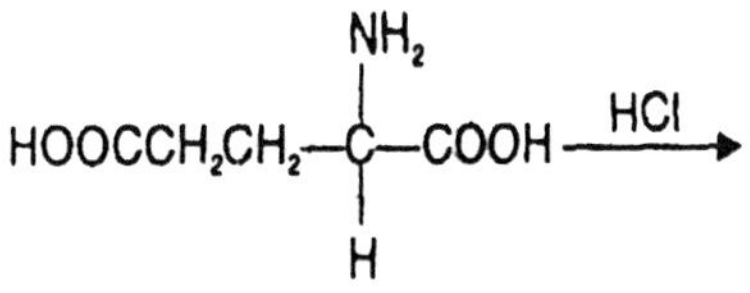

Glutamic Acid

Figure 5.8

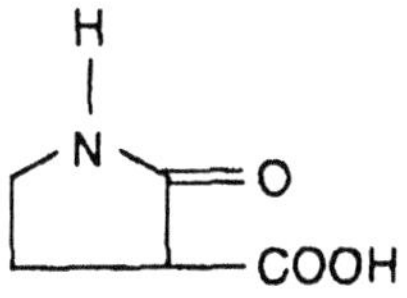

Pyrrolidone Carboxylic Acid

Figure 5.9

till about 40 ml is left. HCl prevents conversion of glutamic acid to pyrrolidone carboxylic acid.

The above 40 ml residue is quantitatively transferred to 50 ml volumetric flask, diluted to volume with water and mixed.

Determination A 25 ml aliquot is transferred to the prepared column, the flow rate is adjusted to 0.5 ml / minute. The column is washed with 10 ml H_2O. The washings are let to pass into resin. 120 ml of 0.8 N HCl is added, maintaining the same flow rate. After all the above acid has passed into resin, 170 ml IN HCl is added, the flow rate is adjusted to 25-30 drops/minute for elution of glutamic acid. The eluate is collected in a 400 ml beaker, nearly neutralized with 50% NaOH and adjusted potentiometrically to pH 7 with 0.1 N NaOH.

25 ml of 37% formalin is neutralized to pH 7 with 0.1 N NaOH and added to the beaker, contents mixed for 10 minutes on a magnetic stirrer and titrated potentiometrically to pH 8.9 with 0.1 N NaOH.

A mixture of 25 ml neutralized formalin + 170 ml 1 N HCl neutralized to pH 7.0 is titrated to pH 8.9. This titre is blank B.

Calculation

% Glutamic acid = (S-B) x N x 0.147 x 100/W, where,

S = ml NaOH used to titrate sample,

B = ml NaOH used to titrate blank,

N = normality of NaOH and

W = weight of sample taken for testing in g.

% MSG = % glutamic acid × 1.15

Before using resin column once again, it is washed with 50 ml of 4N HCl, followed by water. The washings are tested with $AgNO_3$ solution to confirm chloride absence.

HPLC determination of free Glutamic Acid

Free glutamic acid is extracted from the food with 0.02 M KH_2PO_4 with subsequent precipitation of other food components with acetone. Impurities are removed by passing the extract through a C-8 or C-18 reversed-phase solid phase extraction column. The free glutamic acid is derived with phenylisothiocyanate (PITC) and this derivative is separated by HPLC with detection at 254 nm.

The method has been reported by the office of Premarket Approval, US, Food and Drug Administration[13].

In this method, precolumn PITC derivatisation has been adopted because the reagents are easily removed with nitrogen or under vacuum, yielding a relatively clean PITC derivative which has a fairly intense absorbance at 254 nm. Food with/without MSG, natural flavours or hydrolysed vegetable protein can be analysed to determine the level of free glutamic acid present.

Apparatus

1. Liquid chromatograph (Shimodzu, Columbia, MD21046) consisting of two pumps (Model LC-6A) equipped with 100µl injection loop.
2. A variable wave length uv detector (model SPD-6AV), set at 254 nm.
3. A system controller (model ScL-6B)
4. A chromatopak data processor (Model CR-601)
5. A 4.6 x 150nm supelcosil C-18, 5 µm column manufactured by Supelco, INC, Bellefonte, PA 18823.
6. The precolumn consisting of a modular unit with disposable Zorbax RX-C8 filters (MoC-Mod Analytical, INC, Chodds Ford. PA 19317).
7. Flow rate set @ Iml/min.

Reagents

(a) All reagents of A.R. grade.

(b) Deionised water.

(c) The following mobile phase solutions.

(i) ***Solvent A*** 11.45 g of sodium acetate is dissolved in 900 ml water. 0.5 ml of triethyl amine (Aldrich Chem. Co. INC. Milwaukee WI 53233) is added, and the resulting mixture is adjusted to pH 6.4 with acetic acid. 60 ml of acetonitrile is added, the solution is made up to 1 litre with deionised water.

(ii) *Solvent B* 400 ml of deionised water + 600 ml acetonitrile mixed.

Both mobile phases are filtered through an Anorec 0.2 μm filter (Whatman,, INC. Clifton, NJ 07014). The derivatization agents are prepared as follows,

1. The drying agent is prepared by mixing methanol: H_2O : trimethyl aniline :: 1:1:1.
2. The PITC reagent is prepared by mixing 100 ul of PITC (Aldrich C.C.) 100 ul of triethyl amine and 700 ul of methanol. The first two chemicals are toxic, therefore, their safety data sheets must be carefully followed.
3. 0.02 M KH_2PO_4 extraction solution is prepared by dissolving 2.7 g of the chemical in water and making up the volume with water to 1 litre.
4. Stock Standard is prepared by dissolving about 100 mg glutamic acid, accurately weighed, in 100 ml of 0.02 MKH_2PO_4.
5. Working Standards (3/4) are prepared with an L-glutamic acid concentration in the range of 10-60 μg/ml.

Extraction The test food sample is mixed thoroughly, solid foods are ground with pestle and mortar. Typically, a 3 ± 2 gm test portion of the composite food is mixed with 30 ml of 3 (above) and allowed to stand for 15 minutes. 30 ml of acetone is added, the mixture is again allowed to stand for an additional 15 minutes, filtered through a coarse grade filter paper, filtrate diluted to 100 ml with 3 (above). The extract is further diluted to a concentration within the range of the standard curve before proceeding with clean up and derivatization with PITC. Derivatization of an additional aliquot of the extract is necessary if results do not fall within the range of standard curve.

Clean Up A reversed-phased solid phase extraction column (C-8 or C-18) is cleaned twice with 2 ml of methanol and then twice with 2 ml of 3 (above). Two 2 ml aliquots of the extract are passed through the extraction column. The first 2 ml of eluate is discarded and the subsequent 2 ml of eluate is collected for analysis.

Derivatization The derivatization of extracts and working standards can be accomplished manually or semi-automatically by using a Pico-Tag work station (Water Chromatography, Milford, MA 01757). For the semi-automatic procedure, 30 ul of extract is placed in a 6 x 50 mm test tube that has been acid-washed. The test tube is then placed in a specially designed vacuum vial with a resealable polytetrafluoro ethylene (PTFE) closure, and the extract is dried under vacuum in work station. The derivative is dissolved in 300 ul of solvent A. Manual derivatization is performed by the same procedure except that a 50°C water bath and a stream of nitrogen are used to dry the solutions of each step. For quantitation, a 100 ul aliquot of the derivative solution for each extract and working standard is injected onto the C-18 HPLC column (Apparatus for HPLC conditions referred above).

Calculations

A standard curve is prepared by plotting peak height versus concentration for three or four working standards whose concentrations ranged from 10 to 60 ug/ml. The concentration of glutamic acid in the test aliquot is determined by comparing its peak height with the standard curve. The amount of glutamic acid (ppm) in the food is calculated according to the equation.

$$\text{Glutamic acid} = \text{Concentration in test aliquot} \times \text{dilution factor} \times \frac{\text{initial dilution volume (ml)}}{\text{wt. of test portion (g)}}$$

Analysis of Taints and off Flavours

The analysis of taints and off-flavours is only partly different from that of volatile compounds contributing positively to flavour. The differences are caused by the fact that in general, many compounds contribute to the flavour of a product but only a few to taint and off-flavours.

Like flavour compounds, off flavour compounds often occur in very low concentration and may have extremely low threshold values. Investigations in this field are not possible without sophisticated instrumentation. Besides identification of compound(s) causing taints and off-flavours it is very important and imperative to find out whether these have emanated from food constituents or migrated into product or due to serious disturbance of the balance of various compounds constituting the flavour of a food product. Such investigations lead to Hazard Analysis and Critical Control Point (HACCP) systems. The methods used for these purposes are of following nature.

(i) Isolation, concentration and separation of food volatiles.

(ii) Identification and quantification of odour-active chemicals in a complex mixture of volatiles.

Instrumental Analysis

(i) Preparation of the concentrates

(a) ***Extraction*** The most frequently used method to isolate volatile compounds from non-liquid food products is the combined steam distillation extraction (SDE) procedure. This technique has also found many applications in off-flavour studies[14].

The many small alterations made to the original apparatus of Likens *et. al.* are copy rights, however were reported by Moarse *et. al*[15], by Schultz in *loc. Cit.*[16] and by Godefroot[17]. Of these that of Schultz *et. al.* was an important change, making the apparatus usable at reduced pressure for thermolabile compounds. Godefroot modification is a scaled down version of the original SDE apparatus and has been used by several workers like Nunez *et. al.* (1984) for the isolation of the volatiles of grape fruit juice. It was found that the aroma of the concentrate was very similar to that of the juice, more so than that of the concentrates obtained with the original SDE apparatus or by solvent extraction, Heil *et. al.* (1988) for the quantitative analysis of alkylphenols and aromatic thiols in fish.

Soxhlet direct extraction is another way to isolate volatile compounds from a solid product. Fish off flavours have been studied by Paasivirles *et. al.* (1987). Direct extraction is also often used for liquid products, either manually, using a separating funnel, by vigorous magnetic stirring, or in a continuous extraction apparatus.

The extraction solvent is purified by distillation before use to reduce the interference during subsequent analysis.

The yield of polar volatiles increases very strongly for solid sample extraction by stirring in the solvent mixture of $H_2O:CH_2Cl_2:CH_3OH$::4:5:10(v/v) for 3 hours[18].

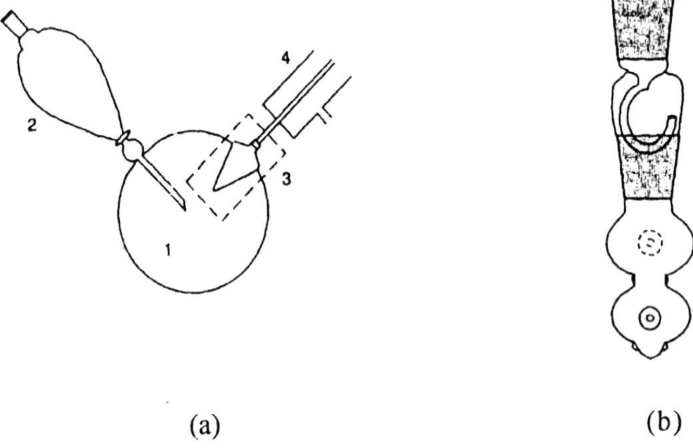

Modification of the jung *et al.*, 1992 apparatus (a) Two necked distillation flask (vol. 500 ml) (1), connected with an addition funnel (2), splash head adaptor (3), and a tube with a water jacket (4); (b) Enlarged display of the splash head adaptor (3).

Figure 5.10

This apparatus is used to separate the volatiles including the extraction solvent from the non-volatile material occurring in the extract[19].

The extract is poured into the distillation flask and frozen in liquid nitrogen. After sublimation of the volatiles on the solvent *ca.* 0.02 Pa for 3 hours, the temperature of the water bath and the water jacket of the tube is increased from 35 to 60°C and the sublimation is continued further for 2-3 hours.

The above apparatus was modified to permit sample dropping into the distillation flask held under vacuum. The splash head prevents the drops spraying into the cooling traps.[20]. The distillation under vacuum is so gentle that even volatile hydroperoxides could be isolated[21].

(c) ***Sample Concentration*** The highly diluted distillates/extracts are concentrated by distilling off the solvent on a Vigreux column until the remaining sample volume is about 1 ml. The sample is transferred as shown below:

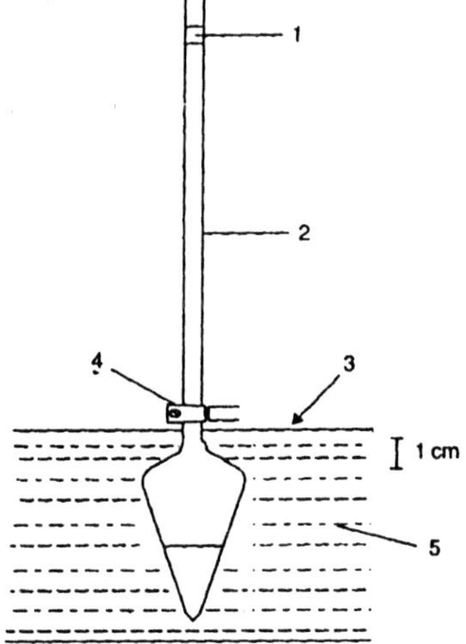

Vessel for the concentration of highly diluted samples (from Dunges, 1979); 1 = condensation zone of the solvent; *2* = tube cooled by air; 3 = thermostatic water bath; 4 = clip; 5 = vessel.

Figure 5.11

Under partial reflux the sample is reduced to less than 35 ml. As the condensation zone acts as a barrier, the loss of the analyte is relatively small so long as the condensation zone has not left the tube[22].

(d) Adsorption

(t) ***Headspace analysis*** In some studies of off flavours, headspace volatiles are collected on adsorbents like Tenax, Chromosorb 105 and charcoal. Further analysis of the collected compounds is possible by heat sorption directly on gas chromatographic column or by extraction with a solvent.

should be prevented in some cases by the addition of salt solutions. In this way concentrations can be obtained that are more representative of the product under investigation. However, sufficient quantity is not always obtained to permit unequivocal identification of trace components. Besides, high molecular and/or polar volatiles are often discriminated against non-polar ones.

For the study of off-flavours in water, an open modification has been developed of the standard water stripping system, using charcoal as the adsorbent. The following advantages are claimed of the open system,

1. Better recoveries for compounds of low volatility and high polarity.
2. Extremely low levels of blanks and simplicity of operation.

Boren *et.al.*[23] have reported the use of extraction to desorb the compounds from the charcoal and compound the efficiency of seven solvents and four solvent mixtures for the extraction of adsorbed compounds from charcoal filters. These desorption studies were performed by adding to the filter a mixture consisting of 50 mg each of 24 compounds widely differing in polarity. Some of their results are shown in Table 5.5.

Table 5.5:
Effect of extraction solvents on the recovery of adsorbed compounds on charcoal (adapted from Boren et. al., 1985)[23]

Compound	*Recovery (%)*				
	Carbon disulphide	*Diethyl ether*	*Methylene dichloride*	*Acetone-carbon disulphide (10:90)*	*Methanol-benzene-carbon-disulphide (5:30:65)*
Butyl acetate	55	62	63	61	65
Chlorobenzene	60	9	39	56	62
1-Clorohexane	57	61	60	61	60
1-Hexanol	43	57	67	62	66

Compound	*Recovery (%)*				
	Carbon disulphide	*Diethyl ether*	*Methylene dichloride*	*Acetone-carbon disulphide (10:90)*	*Methanol-benzene-carbon-disulphide (5:30:65)*
Anisole	62	7	48	62	62
Phenol	3	12	4	52	52
1, 4-Dichlorobenzene	60	2	20	60	60
n-Decane	62	53	52	63	59
Acetophenone	44	2	38	55	60
1-Chlorooctane	63	52	63	63	63
Guaiacol	36	4	39	52	58
1-Octanol	45	47	62	63	57
2, 6-Dimethylphenol	29	6	26	54	53
Napthalene	27	n.d.	3	27	46
1-Chlorodecane	65	37	63	64	65
1-Decanol	49	28	68	59	67
Methyl decanoate	66	33	68	65	65
2-Methoxynaphthalene	18	n.d.	2	21	35
1-Chlorodecane	67	45	62	64	63
Methyl dodecanoate	71	14	67	62	62
1-Chlorotetradecane	62	6	67	60	56
Anthracene	n.d.	n.d.	n.d.	1	4
1-Chlorohexadecane	23	n.d.	49	49	53
1-Chlorooctadecane	34	1	28	37	45
Mean	47	22	44	53	56

Note:— n.d. = not detected

The methanol : benzene : CS_2 :: 5:30;65 was found to be the most effective. When the solvent peaks of this mixture interfere with the subsequent GC analysis, Boren[23] recommended the use of CS_2 with 10% acetone.

(ii) ***Direct adsorption from water*** For the adsorption of traces of volatile components (< 1 ug/kg) in water, XAD-2 use as an adsorbent is suitable in comparison to headspace stripping and direct extraction methods. XAD-2 permits processing of large volumes. However, long-chain hydrocarbons and polar phenols are poorly adsorbed.

Comparison of techniques The stripping technique is less time consuming and very convenient to use, in addition, the absence of high-molecular-weight-compounds in the stripping extracts permits the use of the on-column injection technique, without shortening the life of the GC capillary column. According to Lundgren *et al.*[24], the stripping technique should be preferred if it gives a satisfactory enrichment of off-flavour compounds. Wigilus *et. al*[25], found XAD-2 method and liquid-liquid extraction are complimentary. In the XAD extracts, compounds absent in the concentrates obtained by CH_2Cl_2 extraction, were found and vice versa.

(ii) ***Separation*** In off-flavour studies most of the separations are performed by high-resolution GC (HRGC). Occasionally HPLC is applied. The recent development, specifically relevant to the study of off-flavours are,

(a) HRGC

Analysis of collected headspace samples Collected compounds can be desorbed by solvents, or can be directly injected onto a GC column by heat for which apparatus are commercially available. To prevent peak-broadening, the thermally desorbed volatiles are condensed on a cooled capillary and then transferred to the GC column by heating this capillary. It is also possible to use these capillaries to check by sensory evaluation whether indeed the compound(s) causing the off flavour has (have) been collected.

2-dimensional GC This improves separative power of a GC system elegantly by using two columns in series, a pre-column and an analytical column. Fractions of interest of the first column can be introduced on-line onto the second column and further separated into individual compounds. In one column switching system two ovens are needed in place of one. In the other system a slit is introduced, either between the pre-column and the detector, or between the analytical column and the detector. This enables the selection of compounds contributing to an off-flavour.

Compounds trapping Generally the pre-concentration of the relevant fraction in the chromatogram by trapping that fraction in a cooled capillary is necessary. Direct injection of an off-flavour concentrate is occasionally possible. Jetten[26] developed a method that enabled the injection of the contents of the trap (40 cm x 0.32 mm (i.d.) fused silica capillary column coated with a chemically bonded stationary phase) directly on a GC-MS. The capillary can be mounted in the GC between the injector and the capillary column using a zero dead volume union. Injection on a column with another stationary phase, to enable further separation followed by GC-sniffing, is also possible. Recoveries are between 80 and 97%. The method has the advantage over that of Lundgren *et. al.*[27] in that no solvents are used to remove the compound(s) from the trap.

Pretreated PTFE trap with 3 ul of CH_2 $C1_2$ gave recoveries between 68 and 83%. The relative low yields have been attributed to incomplete recovery of solvent from the trap. Flexibility of fused silica capillaries has its own advantage.

(b) ***HPLC*** It has not found many applications in off-flavour studies. One of the main reason is that it is impossible with the usual eluents to evaluate by sensory analysis the contribution of separated compounds/fractions to an off-flavour of a product. The elements are either toxic, or totally obscure the observation of flavour compounds in a fraction. An element merits selection for its non-toxicity, insignificant flavour and still enables sufficient separation.

HPLC or LC-GC combination can be used to isolate relevant fractions and thus concentrate off-flavour compounds when the identity of the compounds is known. HPLC is also used to qualify known off flavour compounds like chlorophenol. The sensitivity of the detector should be sufficiently low to enable the measurement of concentration below that of the odour threshold value of the compound under investigation as suggested by Whitefield[28].

(c) ***Identification*** Most workers in the field realise that the identity of the compound cannot be decided upon the basis of the measurement of one or two retention times on a

chromatographic column. Additional properties have to be determined, such as mass and/or IR and/or NMR spectra[29].

In off-flavour studies, the comparison of the flavour properties of an unknown and a reference compound could also be important, therefore, reference compounds should be available to measure retention data (preferably RI) and spectra and flavour properties. Except in particular cases, when it is crucial to know with certainty that the identity of a compound is correct it is impossible, in daily routine, to keep up this extremely high standard. Deeming MS and IR spectra as indispensable pieces of information, the reliability of identification can be seen through Table 5.6.

Table 5.6
Rough classification of the reliability of identifications using some major identification methods (from Belz, 1981)

Spectroscopic evidence	*Confirmation by chromatography* [a, b] *Number of retention values* *None*	*One*	*Two*[c]
IR without reference spectrum		-	- +
MS without reference spectrum		-	+ +
IR with reference spectrum	+, ++[d]	++	++
MS with reference spectrum	+ ++	++	++
IR and MS with both reference spectra	++	++	++

a GC, TLC or LC with reference values

b = not reliable; + = moderately reliable; ++ = very reliable

c = GC, polar and apolar stationary phase; thin-layer chromatography (TLC) and liquid chromatography (LC), two different systems

d ++ = first members of a homologous series; + = higher members of a homologous series

Mass spectrometry The GC-MS combination is the most popular technique for identification of volatile compounds in foods and beverages. The latest applications in off-flavour studies are selected-ion-monitoring mass spectrometry (SIM-MS). In this technique the MS is equipped with a peak selector, which instructs the instrument to record selected ions characteristic of the compound of interest. Recording of mass fragmentograms after the analysis is impossible like done by total ion mass spectrometry. Its use is only when one is looking for a compound of known identity. F.R. Sharpe and C.G. Chappell[30] have outlined some advantages of the SIM technique over full scanning.

(i) SIM can improve sensitivity since scanning time can be focused specifically upon ions of interest, and is not spread over wide range. This yields a better signal to noise ratio. A 1000 fold increase in sensitivity can be attained thereby reaching the picogram range[31].

(ii) SIM can improve accuracy and reproducibility for quantitation purposes, and

(iii) With GC-MS, SIM can filter out interfering peaks which may co-elute on the GC column but only if the ion being monitored is absent from the spectrum of the interfering substance.

Whitefield *et. al*[32] could detect chloroanisoles in foods and packaging materials upto 0.01 mg/kg which is sufficiently low (the taste threshold of 2, 4, 6 trichloroanisole in dried fruit is 0.2 mg/kg, for example) T.P. Heil *et. al.*[33] have reported on the basis of a 20 g initial sample, a quantitative detection with SIM-MS of alkyl phenols and aromatic thiols down to a concentration of 0.5 mg/kg.

Sensory methods The only way to select the compounds responsible for an off-flavour is the use of sensory techniques. This implies that studies of off-flavours can only be started when persons are available who can actually detect it. This means that panel members who are trained to detect and describe odours are asked to compare informally an off-flavoured product with a reference product and the same person(s) are asked during subsequent study of the relevant off-flavoured product.

Control of concentration procedures The same panel members are expected to control the isolation and concentration procedures to yield a concentrate with the relevant off-flavour. When collected head space samples are heat-desorbed in cooled gas capillaries (ii)(a) the contents of these capillaries can be used for sensory assessment. When the concentrate is a solvent extract these sensory tests are sometimes difficult because of interference by residual solvent in the concentrate. The use of paper odour strips can partly overcome this problem but the technique depends on persons with experience in this type of assessment.

Selection of off-flavour compounds

(i) ***GC-Olfactometry (GCO)*** The first step in the characterization of odour active chemicals in a complex mixture of volatiles is to localize them in the GC by GCO.

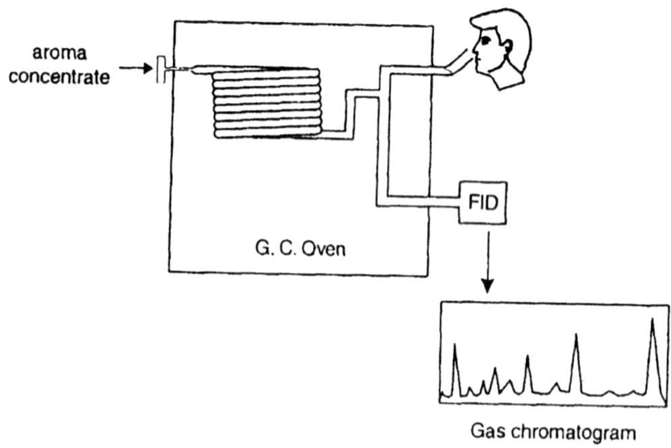

Figure 5.12

As shown above, the effluent of a GC capillary is split 1:1 (by volume) into an FID and sniffing part. The splitter is flushed with the carrier gas helium for accelerating the split flow to about 10 ml/minute as suggested by Blank *et al*.[34]. At the beginning of the chromatogram, compounds leave the column every 5-10 seconds, so in this period particularly, fast translations of sensory impressions into description have to be made. The odour descriptions are noted on a form with a preprinted time scale. Preferably, the assessor should not see the chromatogram.

Sniffing should not last more than 30 minutes, otherwise olfactory fatigue distorts the results. Some workers avoid using the sniffing technique for fear of adverse effects like nasal cancer by the sniffed chemicals like chlorophenol.

GC-sniffing results in an *aromagram,* a gas chromatogram in which all odours perceived are indicated on a time basis by atleast two experts who smell both the eluent of the off-flavoured product, and that of a reference product without the off-flavour. After comparing the results only those peaks that may contribute to the off flavour are selected.

(ii) ***Dilution analysis*** When an off-flavour is caused by contaminants that are usually absent in foods, then GCO runs on two capillary columns of different polarity which at least might be enough to localize this compound in gas chromatogram. When the off-flavour is the result of imbalance of odourants, which are common constituents of foods, their GCO of a dilution series is necessary as the number of odourants detectable by GCO depends not only on the odour thresholds of the volatiles but also on parameters that are selected arbitrarily[35]. The parameters are the amount of food sampled, the dilution of the volatile fraction by the solvent, and the sample size analysed by GC. Consequently, just one run of GCO is usually insufficient to distinguish between the potent odourants that contribute strongly to the flavour of off-flavour and those odourants that are only components of the back ground flavour and are insignificant said LJ. Van Gemert during the workshop "GC-Olfactometry" in 1994.

Two variations of the dilution technique have been developed[36,37]

Both methods are bioassays for odour activity in GC effluents where an extract obtained from the food is diluted with the solvent, diethyl ether usually as a series of 1:1 or 1:2 dilutions, and each dilution is analysed by GCO. This procedure is performed until no odorants are perceivable by GCO. During the runs the retention time of each odour detection is recorded along with a description of flavour.

In Aroma Extract Dilution Analysis (AEDA), the result is referred as a flavour dilution, FD-factor, which is the ratio of the concentration of the odourant in the initial extract to its concentration in the most dilute extract in which odour is detected by GCO. Obviously, the FD-factor is a relative measure and is proportional to the odour activity value, (OAV)- the ratio of concentration to odour threshold of the compound in air.

Charm analysis constructs chromatographic peaks similar to mere produced by chemical detectors in which the peak areas are proportional to the amount of chemical in the extract. The primary difference between the two methods is that charm analysis measures the dilution value over the entire time the compounds elute, whereas AEDA simply determines the maximum dilution value detected.

Performance of the dilution experiment is obligatory using a polar phase like free fatty acid phase, FFAP as well as a non-polar stationary phase like silicone rubber-SE54, because some odourants are better resolved and less adsorbed by one or the other of the two types of GC capillary columns.

H. Maarse and H. Grosch suggest that AEDA on one of these capillaries should be performed by a panelist within 2 days as follows:

1st day: Original extract followed by dilution 2^2 - 2^4 - 2^8 -, etc.

2nd day: dilution 2^1 - 2^3 - 2^5 -, etc.

The odourants contributing to an off-flavour are localized in GC by a comparative AEDA of fresh and deteriorated samples. Clear results are obtained when the corresponding dilution of both samples are analysed by GCO in parallel.

A graphic representation of AEDA can be obtained by plotting the FD value versus the retention time, retention index is better. The result is denoted as *FD-chromatogram.*

AEDA and charm analysis are limited to odourants with a higher boiling point than the solvent used for the extraction and dilution steps. Besides odourants boiling in the same range as the extraction solvent are partially lost during the concentration of the extract by distilling off the solvent. The way out is that AEDA is completed by GCO of decreasing headspace samples. This method, GCOH consists of putting sample A from fresh food and B from the same type of stored food, 50 g each into a vessel, 240 ml each, which is sealed with a septum, and then held in a water both at 40°C. After 1 hour and during a period 1.5 hours at the most, the headspace volumes detailed are withdrawn with a gas tight syringe and then analysed by a GC connected to a purge and trap system. Each injected sample is concentrated in a trap by cryofocusing at 110°C

To start the GC run, the trap is heated very rapidly at 200°C and the headspace sample is flushed with helium on the capillary column. The effluent which is split, 1:1 by volume into an FID and a sniffing part, is examined by sniffing.

A drawback of GCOH is that patent odourants present in two concentrations in headspace samples are not detected by FID and/ or MS for instance. This disadvantage can be overcome when all odourants present in the food extract are first screened by AEDA and then identified. As the extract contains the odourants in much higher concentration, the identification of trace components is easier than in static headspace samples. However, a fraction of these odourants show OAV greater than unity also in GCOH. As the chromatographic and sensory properties of these odourants are known from the preceding AEDA they are quite easy to be identified during GCOH.

The dilution techniques are screening procedures for patent odourants as the results are not corrected for losses during isolation and GC. For indication of compounds factually contributing to the aroma of a particular food and revealed by the dilution procedure, quantification of the odourants with high FD-factors and calculation of their OAV are the next steps in the analytical procedure.

Confirmation After identification and quantification of relevant compounds sometimes it becomes necessary to confirm the contribution of these compounds to the off flavour. No confirmation is needed when one compound is found to have the same characteristic flavour as the off flavoured product. In the final check the compounds are added to the reference product and the panel members are asked to establish whether the flavour is similar or • equal to that of the off flavour product.

Liquids do not create any problem during preparation of a homogeneous mixture of the compound in the reference product. Raw eaten fruits and vegetables are converted into a puree/juice. Dried food together with a filter paper impregnated with a known amount of the compound under investigation is agitated, by placing it on a pair of mechanically driven rollers, in a 5 litre cylindrical sealed glass jar. The quantity of the incorporated compound is calculated by reweighing the filter paper after agitation period (18 ± 6 hours) depending on the volatility of the compound. Processed foods are first ground, minced or blended with a measured quantity of the off flavour compound. Small portions are then rapidly cooked, preferably in a microwave oven, for a few minutes before sensory assessment.

The way off-flavour studies are carried out in practice and to give an insight in the choice of analytical procedure suitable for the type of off-flavour and on the matrix as well as the availability of the equipment, MJ. Saxby has described various techniques applied in his book[39]. Saxby concludes.

"AEDA in combination with GCOH are the most appropriate methods for the identification of substances causing off flavours. However, the simplifications implicit in these methods have to be corrected for by quantification of the levels of these substances and by the calculation of OAVs. Odorants showing much higher OAVs in the off-flavoured compounds with the perfect food sample are suitable as indicator substances for the objective determination of the flavour difference. However, in case of contaminants, originating, for example, from the package material or the environment, a simpler procedure might be useful to solve the off-flavour problem".

References

1. Nadim A. Shaath and Peter Griffin M., in their presentation at the 5th International Flavour Conference at Porto Kanas, Greece 1-3 July 1987.
2. Gurdeep Singh, Kapoor I.P. and Modi Mannon S, *Indian Perfumer,* 1995, 39 (4), pp 149-153.
3. Jennings W., *Gas Chromatography with glass capillary columns,* 1980, 2nd Edn., Academic Press, New York.
4. Frigerio A. in *Essential Aspects of mass spectrometry,* Spectrum Publication, Flushing, N.Y. 1974, pp. 93-110.
5. Gopalkrishnan M. and Narayanan C.S., *Proceedings of the VII PAFAI Seminar,* 1-2 Feb. 1985, Madras, pp 77.
6. Madhusudana Rao J., *Indian Perfumer,* 1988, 32(1), pp 109-112.

7. IOFI, *Z. Lebensm Unters Forsch, 7991, pp 192, 530:*

8. Davies N.W.J. *Chromatogr,* 1990, 1, 503.

9. Biedermann M. and Grole *K,J.H.R.C.,* 1991, 14, 558.

10. Papas A.N. and Delaney M.F., *Anal. Chem,* 1987, pp 54A, 59.

11. Kenneth Farrell T, *Spices, Condiments, and Seasonings 2nd* edition, Van Nastrand Reinhold. 1990, pp 267-269.

12. *Symposium on safety and usefulness of Glutamate as a flavour enhancer: current state of knowledge,* February 15 and 16, 1997. Brindavan Gardens, Mysore.

13. *Food additives and contaminants,* 1995, 12(1), pp 21-29.

14. Saxby M J. *Taints in food stuffs, detection and identification, FMF Rev,* 1973, pp 8, 11-26.

15. *J. Agr. Food Chem.* 1970 (18), pp 1095-1101

16. Schultz *loc. cit.* 1977(25), pp 446-449

17. Godefroot in J. *Chromatogr,* 1981,203, pp 325-335.

18. Semmelroch P. *et. al.* in *Flavour Fragrance Journal,* 1995, 10, pp 1-7.

19. Sen *et. al.J. Agri. Food Chem.,* 1991 (39), pp 757-759.

20. Jung *et al., Lebensum. Wiss, Technol.* 1992(25), pp 55-60.

21. Guth H. and Grosch W., *Lebensum. Wiss. Technol.,* 1990 (23), pp 59-65.

22. Diinges W., *PrechromatographicMicro Methods,* (In German) 1979, Dr. Alfred Hiithig Verlag, Heidelberg.

23. Boren *et.al.J Chromatogr,* 1985 (348), pp 67-78.

24. *Water Sc. Technol,* 1988 (20) (8/9), pp 81-89,

25. *Chromatogr J.,* 1987 (391), pp 169-182,

26. JettenJ.,*J. High Resolution Chromat, Chromatogr Commun,* 1985 (8), pp 696-697.

27. *J. Chromatogr,* 1989(482), pp 23-34

28. Whitefield F.B. and Shaw KJ., *Progress in Flavour Research,* Elsevier Science Publishers. B.V. Amsterdam, 1985, pp 221-238.

29. Marrse H., Introduction in *Volatile Compounds in foods and Beverages,* Marcel Dekker INC, New York, 1991, pp 1-39.
30. Sharpe FR and Chappell *CGJ. Inst. Brew,* 1990(96), pp 381-93.
31. Schreier P. in *Characterization, Production and Application of Food Flavours,* Akadem Verlag, Berlin, 1988, pp 23-42.
32. Whitefield *et.al.J Food Agr.,* 1986 (37), pp 85-96.
33. Heil TP and Lindsay RC,J. *Environ. Sci. Health,* 1988 (B23), pp 475-488.
34. Blank *et al. Z. Lebenson Unters Forsch,* 1989 (189), pp 426-433.
35. Grosch W in *Trends in Food Sci. Technol.,* 1993 (4), pp 68-73.
36. Acree TE, *Charm Analysis* in *Flavour Science - Sensible Principles and Techniques,* American Chemical Society, 1993, pp 1-20.
37. Ulrich F and Grosch W, *Aroma Extract Dilution Analysis (AEDA)* in *Z. Lebensm. Unters. Forsch.* 1987 (184), pp 277-282.
38. Whitefield FB and Show KJ, *Progress in Flavour Research,* Elsvier Science Publishing B.V. Amsterdam, 1985, pp 221-238.
39. *Food taints and off-flavours 2nd Edn,* published by Blackie Academic and Professional, 1996, pp 87-103.

Chapter 6

Sequestering/Buffering Agents

INTRODUCTION

Sequestering agents, according to P.F.A. Rule 70 are

"Substances which prevent adverse effects of metals catalysing the oxidative breakdown of foods forming chelates, thus inhibiting decolourisation, off taste and rancidity".

Lexically, to sequester means to set apart, to seclude, to withdraw, to retire. Metals catalyse the auto-oxidation (Cu, Fe) of oils and fats giving rise to rancidity, off flavour and bad odour. Unsaturated components are more susceptible to autooxidation, saturated are less vulnerable to this oxidative deterioration. To prevent oxidation, unsaturated materials when packed in metal containers are either surrounded by inert gases like nitrogen or sequestering agents are incorporated in them. These sequestering agents form chelates thereby arrest the metals and the resulting aforesaid auto oxidation and its subsequent ill effects.

Chelation means the chemical mechanism through which compound(s) in whom a central metallic ion is joined by ordinary or co-ordinate bonds to two or more non-metallic atoms of the same molecule, so that one or more heterocyclic rings are formed with the central ion as part of each ring. *Chela* means one of the prehensile claws possessed by certain arachnids and crustaceans, as crab and lobster. Like this claw, chelating agents hold the metal ion secluded from the other active ions in a reactive media. For example, to isolate heavy metals, chelation is effected through citric acid which plays a dual role of an acid as well as of a base because it has displaceable hydrogen atoms and a pair of unshared electrons available for coordination simultaneously. Further, the H^+ and OH groups are so spaced that the formed chelate ring is nearly free from strain. Like tartaric acid, citric acid being an (-OH) acid is a better holder of ($^{+++}$) ions in solution at pH7. It is for these reasons that acids like citric are added for chelation of heavy metals like lead, etc.

A general consideration for complex formation, as mentioned by Walton (1966) is that to form a neutral chelated compound with a metal ion, an organic molecule must have a dual character. On the one hand this molecule should be acidic in nature, i.e. it must have displaceable hydrogen atom and on the other hand it must be basic also, i.e. it must have a pair of unshared electrons available for coordination. Further the acidic and basic groups must be so spaced that chelate ring when formed is completely free from strain or nearly so. The (-OH) group is a universal complex former by virtue of its dipole moment and unshared electrons. It can compete with the solvent molecules only if chelation is possible.

Chelating agents which chelate metallic ions such as copper and iron, promote lipid oxidation through a catalytic action. The chelators are sometimes referred to as synergists since they greatly enhance the action of phenolic antioxidants. It is suggested that the synergist (SH) regenerates the primary antioxidant according to the following reaction mechanism.

$$SH + \mathring{A} \quad \rightarrow \quad AH + \mathring{S}$$

Thus, the choice of permitted antioxidant to use, depends on understanding the reaction mechanism outlined under "Antioxidants". For maximum efficiency, a combination of primary antioxidant, synergist and chelating agent is often used. Moreover, the application of chelators and synergists prevents decomposition of hydroperoxides by their sequestering action on catalytic trace metal, thus retarding the development of rancid off flavours and odours in oils, fats and foods containing lipid P.

The alpha (-OH) acids like citric and tartaric are better to hold tripositive ions in solution at pH7. When a dipositive ion is to be held then pH is not kept at 7, tartaric acid is not used as such but in its K/Na salt form for chelation of (Cu^{++}) in Fehling's solution. The chelate configuration is shown in Fig. 6.1.

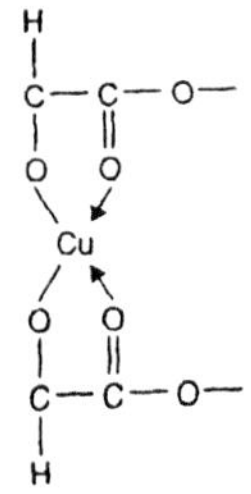

Figure 6.1

Another well known sequestering, buffering and chelating agent is Edectate calcium disodium, also known as Ca Na ethylene diamine tetra acetate, Ca Na EDTA having the following configuration.

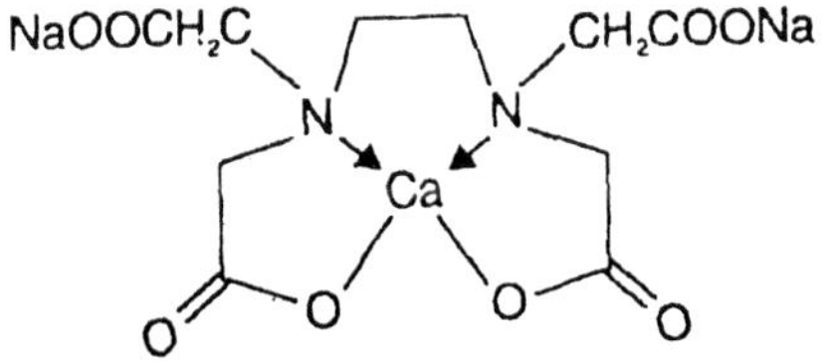

Figure 6.2

It is a tetrahydrate, powder, soluble in water: at 30°C. a 0.1M solution can be prepared, pH7. Practically insoluble in organic solvents. It exchanges its calcium for lead or other heavy metal ions forming water-soluble complexes of the heavy metals.

It is used as a colour retention agent in foods, as a flavouring agent, chelating agent in chemistry and by this virtue as therapeutic antidote for metallic poisonings in veterinary medicine.

Adipic acid (1,4-butanedicarboxylic acid, $HOOC(CH_2)_4$-COOH) is another such example, besides not being hygroscopic is used as an acidulent in baking powders in stead of tartaric acid, cream of tartar and phosphates.

Gluconodelta Lactone

Primarily it is a component of many cleaning compounds because of the sequestering ability of the gluconate radical which remains active in alkali solutions. In dairy industry it is employed to prevent milk stone, in breweries to prevent beer stone. Gluconodelta lactone crystals decompose at 153°C and their sweet taste is different from gluconic acid. Solubility in water 59 g/100 ml, in alcohol about 1 g/100 ml. A freshly prepared 1% aqueous solution has a pH of 3.6 changing to pH 2.5 within 2 hours.

Besides, it is used as a leavening agent in cured meat and meat products.

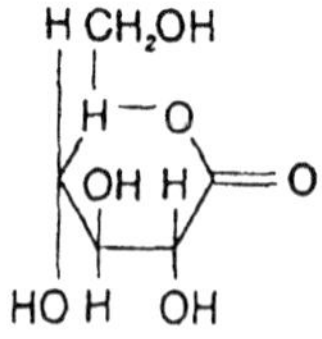

H CH_2OH
H—O
OH H =O
HO H OH

Figure 6.3

Estimation An enzyme procedure has been recommended by ISO and BSI (ISO 4133; BS 4401 part 13).

Reagents

(i) ***Perchloric acid 0.4** M* 17.3 ml perchloric acid 70% m/m is diluted to 500 ml with H_2O.

(ii) ***KOH 2 M*** 56.1 g of KOH is dissolved in water and diluted to 500 ml.

(iii) ***Buffer Solution pH 8.0*** 2.64 g glycylglycine and 0.284 g Mg Cl_2, $6H_2$ O are dissolved in 150 ml water. The pH is adjusted to 8.0 with (ii) above and diluted up to 200 ml with H_2O.

(iv) ***Nicotinamide Adenine Dinucleotide Phosphate (NADP)*** 50 mg NADP $(Na)_2$ salt is dissolved in 5.0 ml water.

(v) ***Adenosine-5-Triphosphate (ATP)*** 250 mg ATP $(Na)_2$ salt + 250 mgm $NaHCO_3$ are dissolved in 5 ml of water.

(vi) ***6-Phosphogluconate De Hydrogenase (6-PGDH)*** Commercial suspension containing 2 mg 6-PGDH/ml from yeast.

(vii) ***Gluconate Kinase (GK)*** Gluconate Kinase suspension containing 1 mg/ml for *E. Coli.*

Procedure

To 50 gm of the prepared sample is added 100 ml cold (0°C) of (i) and homogenised in a laboratory homogeniser. The slurry is transferred to a 100 ml centrifuge tube, centrifuged at 3000 rpm for 10 minutes. The fat layer is removed, the supernatant is decanted through a fluted filter paper into a 200 ml conical flask, the first 10 ml is discarded. 50 ml filtrate is transferred to a 100 ml beaker. pH is adjusted to 10 with (ii) above, the volume made up to 100 ml in a volumetric flask with water. The flask is cooled in ice for 20 minutes. Contents are filtered through a fluted paper, first

10 ml of filtrate is discarded. 25 ml of the filtrate (Vml) is taken in a 250 ml volumetric flask, diluted to the mark with water. Maxium concentration of D (+) gluconate is 400 mg/litre. This is the prepared extract, (pe).

In each of two photometric cells is pipetted 2.5 ml (iii) + 0.01 ml (iv), 0.10 ml (v). Into one of the cells is pipetted 0.20 ml, pe. Into the other 0.20 ml water and 0.05 ml of (vi) is pipetted on to a plastic spatula, contents of one of the cells is mixed and procedure repeated with the other cells contents. Absorbance' of each cell is read at 365 nm against air after five minutes. The cells are retained for the reaction.

A_2 = absorbance of test solution

A_{1B} = absorbance of blank solution.

0.01 ml of (vii) suspension is pipetted onto the plastic spatula. Contents of the cells are mixed alternatively. Absorbance of each cell is read at 365 nm after 10 minutes and every two minutes until a constant rate of increase of absorbance is obtained. Absorbances are plotted against time, the linear part of the curve is extrapolated to zero time.

A_2 = absorbance (T = *zero)* of the test solution

A_{2B} = absorbance (T = zero) of the blank solution

Calculation

$\Delta A = (A_2 - A_1) - (A_{2B} - A_{1B})$

Glucone-delta-lactone %(m/m) in prepared sample

$$\lambda = \frac{15{,}058 \times \Delta A}{V \times M}\left(100 + \frac{M \times m}{100}\right)$$

where V = volume in ml of filtrate to make prepared extract,

M – moisture content of the prepared sample % (m/m),

m = mass in gm of test portion.

Calcium Gluconate

To sample of 200 mg of calcium gluconate, add 3 ml of hydrochloric acid. Bring to boiling pt. and add hot ammonium oxalate solution, followed by dilute ammonium solutions. Heat on water-bath for 1 hour and cool. Transfer precipitates to G4 sintered funnel with the aid of 0.10% w/v ammonium oxalate solution. Wash precipitates with water till free of chloride. Dissolve the precipitates in dilute sulphuric acid, heat to 70°C and titrate with 0.1N potassium permanganate maintaining solution at 70°C throughout titration. Each ml of 0.1N potassium permanganate is equivalent to 0.02242 gm of calcium gluconate, pentahydrate.

BUFFERINGAGENTS

According to the definitions of Bronsted and Lowry, an acid is a proton donor and abase a proton acceptor. Thus an acid-base reaction is supply a protons transfer,

$$HA \quad + \quad B \quad \rightleftharpoons \quad \acute{A} \quad + \quad HB^{+}$$

Acid Base

The products of this reaction form a second acid-base pair; A is called the conjugate base of AA, and HB^+ is the conjugate acid of B. This terminology is used even if the reaction is effectively irreversible; one seldom thinks of molecular hydrogen as an acid, but H~ is certainly a base and so H_2 is its conjugate acid. The charges shown in the equation are purely conventional. An acid or base can be neutral or bear positive or negative charges, and many species can behave as an acid in one reaction and a base in another. Thus the hydrogen sulphate anion, $HOSO_3^-$, can accept a proton to form H_2SO_4 or donate one to form SO_4^- dianion. Likewise the hydrazinium ion, NH_2 NH_3^+, can generate N_2H_4 or $N_2H_6^{2+}$.

The position of an acid-base equilibrium defines the relative strengths of the two acids and two bases present. In the reaction between hydrogen chloride over sodium acetate the equilibrium lies well to the right

$$HCl \quad + \quad AcO^- \quad \rightleftharpoons \quad Cl^- \quad + \quad HO\,Ac.$$

One can regard this as a competition between the two bases, acetate and chloride, for control of proton; the acetate ion wins and is thus the stronger base. Conversely hydrogen chloride is more successful than acetic acid in shedding its proton and is, therefore, the stronger acid.

In this manner a series of acid base equilibria could be used to establish a sequence of acid and base strengths. However, it is more convenient, within practicable limits, to use water both as a solvent and as a reference base, and to relate the strengths of acids to the position of the general equilibrium

$$HA + H_2O \rightleftharpoons H_3O^+ + A^-$$

Solutions are kept sufficiently dilute for activity coefficients to be taken as unity and for water to be regarded as constant. The strength of the acid HA is then measured by the acid dissociation constant or by its negative logarithm.

$$Ka = [H_3O^+]\,[\acute{A}]/[HA]$$

$$pKa = -\log Ka = \log 1/ka = pH - \log [\dot{A}]/[HA]$$

Because a strong acid or base is *fully* ionised in aqueous solution, a very small quantity can make an enormous difference to the activity of a solution. For example, the addition of 0.01 ml = 0.0001 mol of concentrated hydrochloric acid to a litre of water reduces the pH from 7 to 4, i.e. - log 0.0001.

A buffered solution is one that contains comparable concentration of a relatively weak acid and of its conjugate base; it is much less susceptible to such pH changes. Rearranging the above equation, we have

$$pH = \log [A^-] / [HA] + pKa.$$

Thus a solution that contains equal concentrations of sodium acetate and acetic acid will have a pH = pKa of acetic acid (4.75). Suppose $[A^-] = [HA] = 0.01$ M. If we now make the same addition of concentrated hydrochloric acid it will protonate the acetate, reducing $[A^-]$ to 0.0099 and increasing (HA) to 0.0101. The buffer ratio $[A^-] / [HA]$, is barely changed : the pH alters by only $\log (99/101) = -0.01$. Therefore, buffering agents, according to PFA Rule 71 are

"materials used to counter acidic and alkaline changes during storage or processing steps, thus improving the flavour and increasing the stability of foods".

Rule 72. Restrictions on the use of sequestering and buffering agents.—Unless otherwise provided in these rules the sequestering and buffering agents specified column (1) of the Table below, may be used in the groups of food specified in e corresponding entry in column (2) of the said Table, in concentration not exceeding the proportions specified in the corresponding entry in column (3) of ithe said Table:

Table 6.1

S. No.	*Name of sequestering and buffering agents*	*Groups of food*	*Maximum level of use (parts per million) (ppm) (mg/kg)*
(1)	*(2)*	*(3)*	*(4)*
1.	Acetic Acid	(i) Acidulant, buffering and neutralising agent in beverages, soft drinks.	Limited by G.M.P.
		(ii) in canned baby foods	5,000
2.	Adipic acid	Salt substitutes and dietary foods	250
3.	Calcium gluconate	In confections	2,500
4.	Calcium Carbonate	As a neutralizer in number of foods	10,000
5.	Calcium oxide products	As a neutralizer in specified dairy	2,500
6.	Citric acid, Malic acid	Carbonated beverage and as an acidulant in miscellaneous foods.	Limited by G.M.P.
7.	DL Lactic acid (food grade)	As acidulant in miscellaneous foods	Limited by G.M.P.
8.	L(+)Lactic Acid (food grade)	——do——	——do——
9.	Phosphoric acid	Beverages, soft drinks	600
10.	Polyphosphate containing less than phosphate moieties	(a) Processed cheese bread	40,000
		(b) Milk Preparations	4,000
		(c) Cake mixes	10,000
		(d) Protein foods	4,000

(1)	*(2)*	*(3)*	*(4)*
11.	L(+) Tartaric acid	Acidulant	600
12.	Calcium Disodium Ethylene, diamine tetra acetate	(i) Emulsions containing refined vegetableoils, eggs, vinegar, salt, sugar, and spices :	50
		(ii) Salad dressing;	—do—
		(iii) Sandwichspread]	—do—
		(iv) Fat spread	—do—
13.	Fumaric acid	As acidulant in miscellaneous foods	3000 ppm

*Note: DL Lactic acid and L(+) Tartaric acid shall not be added to any food meant for children below 12 months (The lactic acid shall also conform to the specification laid down by the Bureau of Indian standards).

Chapter 7

Emulsifying, Stabilising and Antifoaming Agents

FOOD EMULSIFIERS

Food products are generally multiphase systems in which solids, liquids and air are finely distributed during manufacture to give the finished product the desired structure and shelf life. As many of these phases, e.g. oil and water, are not miscible, substances are required to facilitate phase distribution during manufacture and particularly to ensure stability and prevent separation during storage.

Emulsifiers are often used for this purpose. Due to their molecular structure they are deposited at the phase interface and help to stabilize the phases. This is due to their amphiphillic molecular structure shown in Fig. 7.1.

The emulsifier molecule has at least one group with polar affinity (hydrophilic range) and one group with non-polar affinity. With food emulsifiers the lipophillic range consists almost exclusively of edible fatty acids, while the hydrophilic, water-soluble range can be extremely varied. Apart from glycerol and condensed polyglycerol, there is propylene glycol, sorbitol, sucrose or lactic acid, all of which have hydroxyl groups and form esters with fatty acids. Through a subsequent reaction by the esters with organic acids like acetic, citric or tartaric acid or, in the case of ammonium phosphatides, with phosphoric acid, the hydrophilic component of the emulsifier can be further increased.

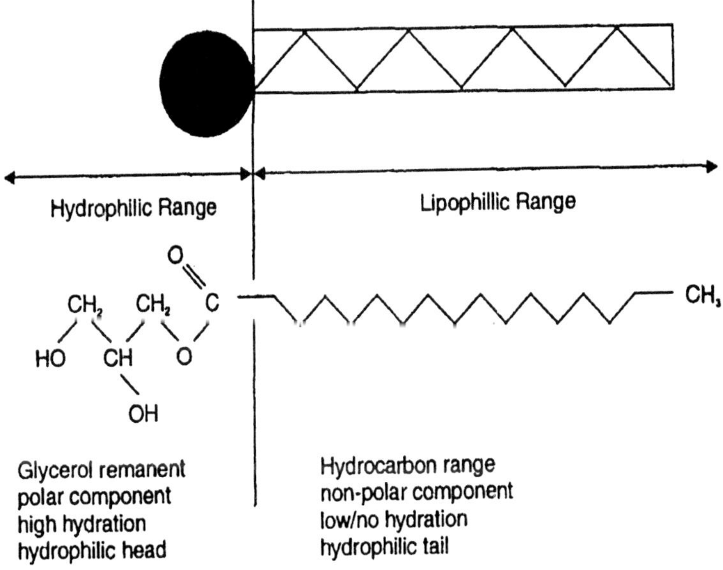

Structure of Emulsifier molecule[1]

Figure 7.1

The hydrophilic effect can also be improved through neutralization of the free acid groups with Na/K/Ca salts. Franz Timmermann has drawn a chart of chemical classification of food emulsifiers which can be seen in Table 7.1.

Table 7.1
Chemical Classification of Food Emulsifiers

Name (1)	*Examples* (2)
A. Fatty Acid Esters	
(a) (i) Monovalent alcohol esters	
(ii) mono-hydroxide acid	Na/Ca stearyl actylate (E481, E482)
(b) (i) Bivalent alcohol esters	
(ii) glycols	1, 2 propylene glycol esters (E 477)
(iii) polyglycols	Polyoxy ethylene (40) stearate (E 431)

Name *(1)*	*Examples* *(2)*
(c) (i) Trivalent alcohol esters	
(ii) glycerol	Mono-and diglyceride (E 471)
(iii) mono-and diglycerides esterified with acid	
(iii) acetic acid	Acetylated monoglyceride (E 472 a)
(iv) lactic acid	Lactic acid monoglyceride (E 472 b)
(v) citric acid	Citric acid monoglyceride (E 472 c)
(vi) tartaric acid	Tartaric acid monoglyceride (E 472 d)
(vii) diacetyl tartaric acid	Diacetyl tartaric acid monoglyceride (E 472 e)
(viii) several edible acids	Mixed acid and tartaric acid monoglyceride (E 472 f)
(ix) oleoresin acids	Glycerine esters from roots (E 445)
(x) phosphoric acid	Ammonium phosphatide (E 442), lecithin (322)
(xi) polyglycerol	Polyglycerine ester (E 475)
(d) Polyhydric polyol esters	
(i) sorbitan esters	Sorbitan esters (E 491-E 495)
(ii) sucrose	Sucrose ester (E 473)
	sucrose acetatisobutyrate (E 444)
(iii) glycerol and sucrose	Sucrose glyceride (E 474)
B. Fatty Acid Salts	Na-K-Mg-Ca salts from edible fatty acids (E 470 a, b)
C. Fatty Alcohol Esters	Stearyl tartarate (E 483)
D. Oxidative Polymerised Soy Oil	Tosom (E 479 b)

Through the presence of hydrophilic and lipophillic groups in the molecule, emulsifiers interact with water and with fats and oils.

The emulsifier is orientated at the phase interface, with the hydrophilic component going towards the water and the lipophillic component towards the oil. This causes a film formation which reduces the interfacial tension.

Depending on the ratio of hydrophilic and lipophillic components, the emulsifier is more readily soluble in one or other of the phases as shown in Fig 7.2.

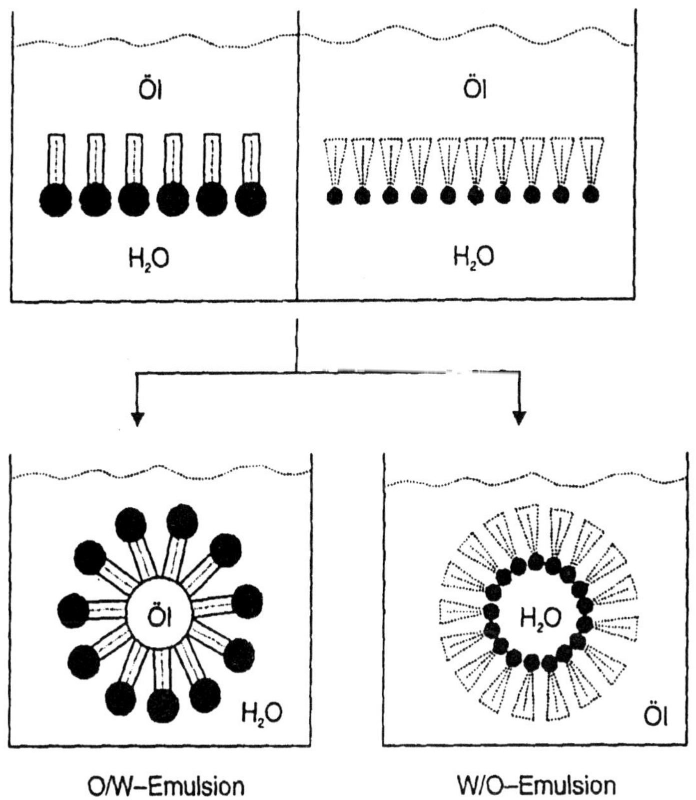

O/W–Emulsion = oil-water emulsion
W/O–Emulsion = water-oil emulsion
Öl = Oil

Emulsifier effect in emulsion formation

Figure 7.2

Polar or water-soluble emulsifiers form an oil-in-water emulsion, while non-polar or oil soluble emulsifiers form a water-in-oil emulsion. One of the main criteria when selecting an emulsifier is thus the HLB, hydrophilic-lipophillic balance, which indicates the relationship between the hydrophilic and lipophillic components. The properties of an emulsifier as a function of HLB can be seen in Table 7.2.

Table 7.2
Emulsifier Properties as Junction of Hydrophillic-Lipophillic Balance (HLB)

HLB Value	*Application*					
1.	1–3 Anti foaming Agent					
2.						
3.		3–6 Water-oil Emulsifier				
4.						
5.						
6.						
7.			7–9 Wetting Agent			
8.				8–18 Oil-water Emulsifier		
9.						
10.						
11.						
12.						
13.					13–15 Washing Detergent	
14.						
15.						15–18 Solubiliser
16.						
17.						
18.						

Besides the above emulsifying properties, emulsifiers are frequently employed to exploit the interaction between fats, proteins and carbohydrates.

Emulsifier—lipid interaction Triglycerides can occur in different crystal forms. When melted fat cools, the alpha form occurs initially, giving the product a waxy structure. Through slow cooling the major beta form occurs. Its microscopy reveals crystalline clusters. It is the most stable form for some fats, other fats can transform into block-shaped beta crystals of 75 ± 25 mm size, imparting the food product an unwelcomed sandy texture. Thus due to different crystalline forms the resulting organolepticity differs. To illustrate, fat-containing whipped desserts, icecream and pastries required alpha or beta crystals for stability. Lipophillic emulsifiers with low HLB value usually inhibit or decelerate the formation of stable beta forms. In margarines, diglycerides delay the crystalline growth of fats; in chocolates sorbitan tristearate slows down the fat bloom.

Emulsifier-Carbohydrate Interaction Amylose can crystallize out of the aqueous helix structure, causing white bread to go stale. The fatty acid remanent in many emulsifiers can have a stabilizing effect by enclosing the helix structure, there by keeping the bread soft far longer. The emulsifier also binds water-soluble starch and improves the texture of starchy products such as noodles or mashed potatoes.

Emulsifier-Protein Interaction The hydrophilic component of emulsifiers can interact with proteins as a result of ionic charges or through hydrogen bridges, there by improving the structure of many food products. In yeast-risen baked goods the gas-retaining capability of the wheat gluten is improved, making far better volume and crumb structure.

Summing up, food emulsifiers are useful not only for their emulsifying properties but also for other effects they produce (Table 7.3).

Table 7.3
Effects of Emulsifiers

Effect	*Food*
Emulsification	Dressings, margarine
Fat crystallisation control	Icecreams, margarine
Foam prevention	Jam
Foaming	Pastries, ice-creams, desserts
Lubrication	Extruded products, separators
Protein interaction	Bread
Solubilization	Flavourings, drinks

Effect	*Food*
Splash protection	Margarine
Starch complexing	Bread, noodles, mashed potatoes
Thickening	Nut pastes
Viscosity reduction	Chocolate
Wetting	Soups

Emulsifying and stabilising agents, as defined under Rule 60 of P.F.A. Rules 1955, are substances which, when added to food, are capable of facilitating a uniform dispersion of oils and fats in aqueous media, or *vice versa,* and/or stabilising such emulsions.

It is a long list, however, under rule 61 of the PFA Act 1954, it has been stipulated that

"no emulsifying or stabilising agent(s) shall be used in any food, except where their use is specifically permitted".

A variety of organic compounds form the group of emulsifiers, stabilizers and thickening agents. Compounds like stearyl tartarate, glycerol esters like glyceryl monostearate, propylene glycol esters, sorbitan esters of fatty acids, cellulose ethers and sodium carboxy methyl cellulose are in use or making food emulsions and to stabilize them. Pectin, alginates, agar, Irish moss, cellulose, carboxyl methyl cellulose, starch and certain gums like guar gum, gum arabic karayagum, gum ghatti, tragacanth gum, locust bean gum, gelatins, etc. are being used as thickening agents.

ALGIN AND ALGINATES IN FOODS

Algin and alginates are extensively used in the preparation of icecreams, in the production of dairy products, such as whipped-cream products, milk powders, milk-mixing products, milk shakes, candies, processed cheese, soft cheese and cheese-creams. They act as stabilizers in cream substitutes and chocolate-milk suspensions. They prevent the formation of ice-crystals during the process of freezing and produce uniform viscosity, good whipping-ability, smooth melting and full development of aromatic components. In food industries like bakery, confectionery and sweetmeats, preparations of various jelly products, meat and sausage, alginates find a wide range of application. They are used in cakes, fruit juice powders, fruit drinks, marmalades, meringues, jams, artificial cherries, Turkish delight, custard and marsh mallows. In sugar-grained confections, they modify the crystal size without affecting flavour. In beverages, such as beet juice and rose wines, they are used for classifying and purifying and also as foam stabilizers. In margarine industry, they serve as emulsifiers and stabilizers as well as protective colloid for fat soluble vitamin A. Deep freezing of fish in alginate gels is an important process, developed in many countries. Triethanolamine alginate finds use as a coating material for cheese, meat and other foods.

Algin and alginates are used in many well known slimming-diets for diabetic and obese persons. They are also used as plasma-substituting colloids.

Algin is sodium alginate, the sodium salt of alginic acid also called sodium poly mannuronate; alto, alman, alloid, allose, kelgin, minus, protanol, sodium alginate. It is a gelling polysaccharide extracted from giant brown sea weed or from horse tail kelp. It is a cream coloured powder, soluble in water, forming a viscous, colloidal solution. It is insoluble in alcohol and hydro-alcoholic solutions in which alcohol content is more than 30% (w/w). It is insoluble in chloroform, ether, in aqueous acid solutions when the pH is below 3.

Detection of alginates in foods The method is more suitable for foods like chocolate, ice-cream and frozen products.

Reagents

(i) $FeCl_3$ - H_2SO_4 reagent 10 g of $FeCl_3$. H_2O is dissolved in about 100 ml of water in each of 2 centrifuge bottles and precipitated as Fe $(OH)_3$ by adding excess of NH_4OH. The precipitate is washed with about 5 portions of water, centrifuged and decanted until little odour of NH_3 remains. The centrifuged precipitate is broken up before washing. The precipitate is dried on steam bath or in oven overnight, broken up and dried again. It is then ground, 500 mg of this powder is placed in 50 ml graduated flask, 50 ml H_2SO_4 is added, contents shaken vigorously and left to settle until it is clear. Some $Fe_2(SO)_3$ appears to stick to sides, but the reagent is ready for use after seven days. Its shelf life is only 3 weeks.

The reagent is checked before use by dissolving 3 ± 2 mg of alginate in water containing 5 drops of 0.1N NaOH, 4 volumes of alcohol added to precipitate alginate. It is centrifuged, decanted and dried on a steam bath until alcohol odour is not perceptible. With air current last traces of alcohol are also removed. 3 drops of 0.1 N NaOH are added, contents dissolved with the aid of a glass rod and 2 ml of $FeCl_3$ - H_2SO_4 reagent is added. The solution slowly turns purple, usually within an hour depending upon amount of alginate present but may take longer period. If the solution appears to be turning brown, additional 2 ml of reagent is added, mixed with glass rod and let to stand.

Procedure

Sample containing 15 ± 5 mg of alginate is weighed into a 25 ml centrifuge bottle, water upto 15 ± 5 ml volume is added, the contents dissolved by swirling, pH is adjusted between 8 to 9 with saturated Na_3PO_4 solution, usually 5 drops suffice. About 500 mg of pancreatin + 3 drops of formalin are added, contents shaken vigorously for a minute then kept aside for a minimum period of two hours but not more than 16 hours followed by centrifugation at 12 rpm for about 3 minutes, decantation into 250 ml centrifuge bottle. The residue is discarded. 3 volumes of alcohol added, contents shaken then left for 1 hour. After shaking several times the contents recentrifuged as before, discarding the liquid, 3 g of decolourising charcoal is added, the contents are shaken vigorously for 1 hour on a mechanical shaker, poured directly into folded filter paper, collecting filtrate in a 250 ml centrifuge bottle. If the filtrate is not clear then it is poured back through filter paper several times. In the event of slow filtration it is completed overnight. 4 volumes of alcohol are added to the filtrate, contents shaken and kept for 1 hour or overnight if convenient. After centrifugation, decantation is done, keeping aside the residue. Residue contains alginates, gums and gelatin, it is dried on a steam-bath, using air current if needed until it is free of alcohol. After cooling, 3 drops of 0.1 NaOH are added, by using glass rod the residue is dissolved as completely as possible. 2 ml of $FeCl_3$-H_2SO_4 reagent solution added. If the solution turns to purple it indicates the presence of alginates.

Ghatti gum Indian gum, the gummy exudate from stems of *Anogeissus litifolia.* It is abundant in India and Sri Lanka. Ghatti, i.e. from ghats, because of its ancient mountain transportation. Ghatti gum sold in the US usually has been autoclaved in order to make all the gum water-soluble. It forms a very viscous mucilage, more viscous but less adhesive than acacia. Insoluble in 90% alcohol, $[\alpha]_D^{25}$ is + 42° (dil H_2SO_4). Gum ghatti solutions may be coloured slightly due to traces of pigments remaining in the gum. It does not form a true gel and is used as a substitute for acacia.

Gelatin Gelfoam, puragel, a hetrogeneous mixture of water-soluble proteins of high average molecular weight. Gelatin is not found in native but from collagen q.v. by hydrolytic action. It is obtained by boiling skin, tendons, ligaments, bones, etc. with water. It is colourless or slightly yellow, transparent, brittle, practically odourless, tasteless sheets, flakes, or coarse powder. Gelatin swells up and absorbs 5 to 10 times its weight of water to form a gel in solution below 35-40°C. It is soluble in hot water, glycerol, acetic acid, insoluble in organic solvents and is amphoteric.

Nutritionally, gelatin lacks tryptophan and other amino acids are present in small amounts making it an incomplete protein.

Table 7.4
Principal Chemical Compounds in Thickening Agents

S. No.	*Material*	*Source*	*Principal compounds*
1.	Pectin substances	Fruit	Galacturonic acid
2.	Algin (Na Alginate)	Seaweeds	Mannuronic acid (Na salt)
3.	Irish moss	do	Galactose, galactose-4 sulphate (K and Ca Salts)
4.	Agar	do	Galoctose 6 - Sulphate (Ca, Mg) salts, galactose
5.	Tragacanth	Plant gum	Fructose, D-xylose, arabinose, D-glalactose
6.	CH_3-cellulose	Modified cellulose	CH_3-D-glucose
7.	Starch	Plants	D-glucose

S. No.	Material	Source	Principal compounds
8.	Carboxyl CH_3cellulose	Modified cellulose	Carboxymethyl D-glucose
9.	Locust-bean gum (Carob gum)	Seed Endosperm	Mannose and galactose
10.	Guar gum	do	do
11.	Karaya gum	Plant gum	Galactose, rhamnose, tagatose, acetic and galacturonic acid
12.	Arabica (acacia)	do	D-galactose and arabinose (mixed Ca, Mg and K salts) and galacturonic acid
13.	Ghatti gum	do	Arabinose, galactose
14.	Gelatin	Modified protein	Amino acids

Pectin

It is a polysaccharide substance present in cell walls in all plant tissues which functions as an intercellular cementing material. One of the richest sources of pectin is lemon or orange rind which contains about 30% of this polysaccharide having molecular weight from 20,000 to 40,000.

It occurs as a coarse or fine powder, yellowish-white in colour, practically odourless, and with a mucilaginous taste. Almost completely soluble in 20 parts of water, it forms a viscous solution containing negatively charged, very much hydrated particles. Acid to litmus, insoluble in alcohol or in diluted alcohol, and in other organic solvents.

Dissolves more readily in water, if first moistened with alcohol, glycerol or sugar syrup, or if first mixed with 3 or more parts of sucrose. Stable under mildly acidic conditions, more strongly acidic or basic conditions cause depolymerisation.

Agar

Also known as Agar-agar; gelose; Japan agar; Bengal isinglass, Ceylon isinglass; Chinese isinglass; Japan isinglass; Layor Carang. A polysaccharide complex extracted from agarocytes of algae of the *Rhodophyceae* besides others. It can be separated into a neutral gelling fraction-agarose, and a sulphated non-gelling fraction-agaropectin.

Agar is transparent, odourless, tasteless strips or coarse or fine powder. It is insoluble in cold water, alcohol, slowly soluble in hot water to a mixed solution. A 1% solution forms a stiff jelly on cooling.

Methyl cellulose

Cellulose methylether, methocel, cellothyl, syncelose, bagolax, celevac, cellucon, cethylose, cethytin, cologel, cellumeth, hydrolose, nicel, tearisol, tylose—all these are trade names of methyl cellulose, composed of white granules. It is odourless, tasteless, soluble in water, insoluble in hot water. An aqueous solution is best prepared by dispensing the granules in hot but not boiling water with stirring and chilling to 5°C. The solution is then stable at room temperature. Presence of inorganic salts increases the viscosity. Its solubility is dependent upon the degree of substitution. Commercial product has a methoxyl content of 29% (degree of substitution about 1.8). Clear films are cast from aqueous solution.

Locust bean gum

Also known as Carob gum/flour, johan-nisbrotmehl, arobon, the ground kernel endosperms of tree pods of St. John's bread *Ceratonia siliqua(L).* It consists of proteins such as albumins, globulins, prolamins, gluteline, carbohydrates like reducing sugars, sucrose, dextrins, pentosans, ash, crude fibre and moisture. Also used as coffee, chocolate, cocoa substitute extender.

It is yellow-green in colour, is odourless and tasteless, but acquires a leguminous taste when boiled in water.

Gum tragacanth

Tragacanth is the dried gummy exudation from *Astragalus gummifer* (Labill), white gavan. When mixed with water it gives a soluble fraction, as a hydrosol, called *tragacanthis* which is a complex mixture of polysaccharides containing D-galaturonic acid, other sugars and traces of starch and cellulose. The insoluble fraction swells to a gel and consists of 65 ± 5% bassorin, *q.v.*

It is odourless, having an insipid, mucilaginous taste. Acid reaction, one gram requires 0.9 ml of 0.1 N NaOH for neutralization to phenolphthalein. Viscosity of tragacanth mucilages is reduced by adding acid, alkali and NaCl, particularly if the mucilage is heated. Maximum initial viscosity of solutions is at pH 8; maximum stable viscosity is near pH 5. It forms a deep yellow stringy precipitate when a solution is boiled with a few drops of 10% aqueous $FeCl_3$ solution. A stringy precipitate formed also on heating a solution with Schweitzer reagent. Tragacanth is entirely insoluble in alcohol.

Guargum

Also known as Guar flour, decorpa, jaguar, gum cya-mopsis, burtonite, V-7-E, guarina and glucotard; Guarem. The ground endosperm of *Cyamopsis tetragonolobus (L).* Taub, which is cultivated in India as a livestock feed. The water-soluble fraction (85%) of guar flour is called guaran. It is a free flowing powder, completely soluble in cold and hot water; especially insoluble in oils, greases, hydrocarbons, ketones and esters. Water solutions are tasteless, odourless, non-toxic of a pale, translucent gray colour, and neutral. It is thermostable, has 5 to 8 times the thickening power of starch. Water solution are convertible to a gel by small amounts of borax: Aqueous dispersions are neutral.

Karaya gum

Also known as kadya, katilo, kullo, kuteera, stercu-lia, Indian tragacanth and mucara. The dried exudate of the tree *Sterculia urens,* (Roxb), is found in India, especially in the Gujarat and M.P. region. It is finely ground white powder, faint odour of acetic acid, acid to litmus, absorbs water rapidly to form viscous mucilages at low concentrations. Viscosity decreases on addition of acid or alkali. Colour of the solution lightens in acidic media and darkens in alkaline solution due to the presence of tannins. Gum karaya loses its mucilage forming ability when stored in the dry state, the loss being greater for a powdered material than for the crude gum. Cold storage inhibits this degradation. It is a substitute for gum tragacanth.

Karaya gum occurring in broken irregular pieces having a somewhat crystalline appearance is commercially denoted as "crystal" gum.

Acacia gum

Also known as gum arabica. It is the dried gummy exudation from the stems and branches of *Acacia Senegal, (L)* Willd, or other African species of *Acacia.* According to C L Mentell, *The water-soluble gums, 1947,* Kordofan gum (Hashbe geneina), the gum from *Acacia verek,* (Guill and Perr) from plantations in the Kordofan Province, Sudan is considered the best commercial variety. Grades of Kordofan gum which are clean, white, sun bleached and tasteless are preferred for food preparation and pharmaceuticals. There is a close relationship between colour and flavour due to the presence of tannins. DMW Anderson[2] has reviewed its use as food additive in *Food Addit. Contam,* 1986,3,225-230.

Occurs in spheroidal tears upto 32 mm in diameter. It also occurs as flakes and powder. Solutions of gums from *A. verek,* are laevo rotatory, other acacia species are dextro rotatory. Sp. gravity 1.42 ± 0.07 and samples dried at 100°C are heavier. Moisture content usually varies $14 \pm 1\%$. Material containing less than 12% moisture chips easily and produces dust during transportation. It is insoluble in alcohol, almost completely soluble in twice its weight of water. 100 g of a saturated solution contains 37 g at 25°C, 38 g at 50°C, 40 g at 90°C. Aqueous solution is acidic to litmus. Also soluble in glycerol and in propylene glycol, but prolonged heating (several days may be needed for complete solution (about 5%).

IDENTIFICATION OF STABILISING AGENTS AND GUMS

Under Rule 61(C) of the PFA Act, the following emulsifying and stabilising agents may be added to fruit products,

(i) pectin

(ii) alginic acid, its Na/Ca salts and propylene glycol alginate

Extraction of gums from fruit and vegetable products

The solid sample is ground in a blender whence 50 g is weighed. Solid or semisolid samples are slurried by the addition of an equal quantity of water, filtered or centrifuged. In the later case the supernatant liquid is decanted. To about 10 ml of the decanted sample 20% tannic acid solution is added and to another 10 ml of the sample 40% trichloroacetic acid is added. The volume of the two precipitates is compared. Tannic acid precipitates only proteins and alginic acid. Depending upon the results of the tests, the remaining supernatant liquid is treated with either tannic acid or trichloroacetic acid for complete precipitation, then kept undisturbed for 15 minutes and finally centrifuged for 5 to 10 minutes. A portion of this supernatant liquid is rechecked by adding the above precipitant so that no further precipitation takes place. The centrifuged contents are filtered. To the filtrate, 5 volumes of alcohol are added with constant stirring, the solution is kept aside upto 10 minutes. It is made alkaline by dropwise addition of ammonia, again kept aside for 5 minutes, and finally cone HCl is added with vigorous stirring for mixing.

A precipitate at this time indicates the presence of gums, starches, dextrins, etc.

The above mixture is kept undisturbed for overnight. Next morning it is filtered, the precipitate is washed thoroughly with alcohol, the residue is dissolved in 30ml of boiling water, boiled if needed for dissolution. This solution (say A) is used for identification.

Identification of gums

400 ± 100 mg of the test material is wetted with about 2 ml of 95% alcohol and 50 ml water. The solid material is suspended in water by shaking or stirring, heated when the sample dissolves. Heating is discontinued, otherwise the contents are held at 90 ± 5°C for 15 minutes (B).

Group I

4 ± 1 ml aliquot of B is treated with 0.2 volumes of 0.25 M $CaCl_2$. A gelatinous precipitation or gel indicates alginates or deesterified pectin. If no reaction is apparent with $CaCl_2$ alone then 1 volume of 3 N NH_4OH is added. Slow formation of a gel or gelatinous precipitate indicates pectin.

If the above test is positive, a fresh 4 ± 1 ml aliquot of the sample is mixed with 3 N NaOH. The reaction is watched then contents are heated in a boiling water-bath for 10 minutes. Immediate formation in the cold of a gelatinous or flocculent precipitate indicates either pectin or deesterified pectin. Absence of precipitate indicates alginates.

All the three mixtures become yellow on heating but the precipitate with pectic substances do not dissolve.

Group II

If Group I is absent then 4 ± 1 ml aliquot of the sample is mixed with one volume of saturated $Ba(OH)_2$. Like before, the reaction is observed in the cold, then contents are heated in a boiling-water bath for 10 minutes. Formulation in the cold of a non-setting, almost opaque gelatinous precipitate indicates Irish moss. A small amount of flocculent precipitate or cloudiness in the cold and definite lemon yellow colour on heating indicates gum tragacanth. If the colour changes during heating to yellow-green-gray, it indicates agar.

If the mixture is cloudy or forms a gel on heating but becomes clear on cooling, CH_3 cellulose is indicated. An opaque flocculent precipitate which may tend to redispense on heating and again precipitates on cooling indicates starch.

Precipitates which disappear when the $Ba(OH_2)$, is thoroughly mixed with the sample may be ignored at this stage.

When all the above tests have been found negative a fresh 4 ± 1 ml of aliquot of the sample is mixed with one volume of saturated $Ba(OH)_2$. Immediate precipitation and examination after 5 minutes of keeping aside is interpreted as,

(i) A voluminous, opaque, stringly precipitate which tends to clot indicates locust bean gum. Shaking makes this precipitate flocculent.

(ii) An opaque flocculent precipitate which becomes sloppy and is not voluminous indicates gum karaya.

Group III

Absence of all the above ingredients indicates that the test sample may be gum arabic/ghatti/gelatin. Another fresh aliquot of the sample is mixed with 1 ml of basic lead acetate solution. Immediate formation of a voluminous opaque precipitate indicates gum arabic.

If the precipitate is small in amount and flocculent in appearance or no precipitate then 1 ml of 3 N NH_4OH is also added. A voluminous opaque precipitate indicates gum ghatti. No precipitation suggests presence of gelatin.

Reactions Useful for Characterisation of Gums

(i) The reaction with Stock's acid, $Hg(NO_3)_2$ illustrate the effects of low pH on precipitation of heavy metal salts of the polysaccharide acids. An excess of the reagent makes the solutions strongly acidic and thus the weakly dissociated acid redisperse. Alginic acid and pectic acid are insoluble hence are not dissolved by an excess of Stock's reagent. Gelatin gives a pronounced precipitation reaction only with those gums having anionic components. The precipitates are formed only if the pH of the mixture is below the isoelectric point of the protein and may be more correctly called concentrates. These are usually dispersed by a few drops of mineral acid or dil. NH_4OH.

(ii) $(NH_4)_2SO_4$ gives pronounced precipitation tests with several of the gums but not with alginates, pectin, tragacanth, karaya, arabic or ghatti, each of which contains uronic acid components.

(iii) Precipitation by alcohol is characteristic of gums enabling their identification.

Confirmatory Tests

(a) ***Alginates and deesterifiedpectins*** 3.5 ± 0.5 ml of sample solution + 0.2 ml of 3 N HCl (or other mineral acid). A gelatinous precipitate is confirmation.

(b) ***Irish moss*** 1 ml of sample solution + 3 drops of 0.5% methylene blue solution in water. Precipitation of purple stained fibres is confirmation.

(c) ***CH_3 Cellulose*** 5 ml of sample solution + 25 ml of 95% alcohol + 3 drops of NaCl. No precipitate is confirmation.

(d) ***Agar*** 5 ml of sample solution + alcohol-precipitate, stain with iodine tincture- a blue colour confirms agar as well as starch presence. To confirm later's presence, 1 ml of sample + 2 drops of iodine tincture - a blue/purple colour is deemed confirmation because same sample of gum tragacanth may give a faint blue test which may be mistaken for starch.

(e) ***Carboxymethyl cellulose***

(i) 5 ml sample solution + 2 ml of 1M $CuSO_4$ solution—an opaque, slightly bluish clotted precipitate confirms its presence.

(ii) 5 ml sample solution + Uranyl Zn acetate-yellow precipitate forms in the presence of CM cellulose.

(f) ***Locust Bean Gum*** 4 ± 1 ml of sample solution + 1 ml of 4% borax-gelatinization means presence of this bean gum as well as guar gum.

(g) ***Karaya gum*** 5 ml of sample solution + alcohol-precipitate, stain it with ruthenium red-considerable pink stained swelling is confirmation.

(h) ***Gelatin*** 3 drops of sample solution + 2 ml of saturated picric acid-a fine yellow precipitate confirms gelatin.

Dimethyl polysiloxane

Also known as Dimethicone, Simethi-cone, Dimeticon, Aeropax, Antifoam A, Baros, Bicolon, Ceolat, Delesan, Endo-paractol, Infacol, Lefax, Meteorex, MS Antifoam M, Mylicon, Mylocon, Ovol, Phasil, Phazyme, Polysilon, Sab Simplex, Silain, Siligaz, Spreadasil silicon spray (Fig. 7.3)

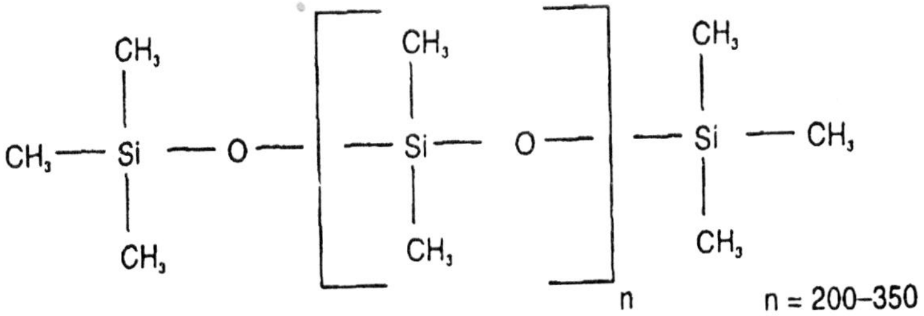

$$CH_3-\underset{CH_3}{\overset{CH_3}{Si}}-O-\left[-\underset{CH_3}{\overset{CH_3}{Si}}-O-\right]_n-\underset{CH_3}{\overset{CH_3}{Si}}-CH_3 \qquad n = 200\text{–}350$$

Figure 7.3

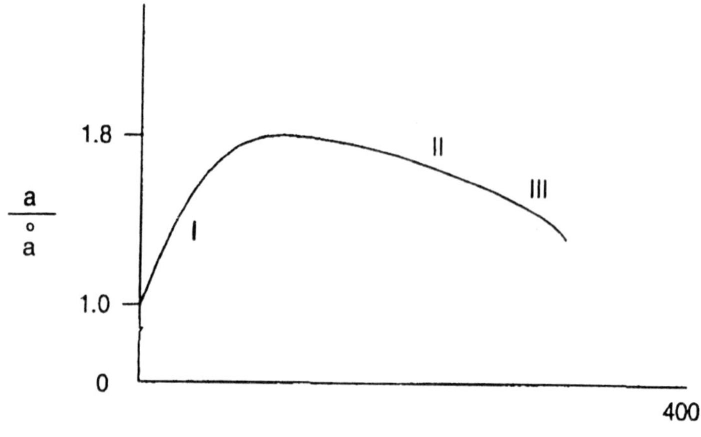

Figure 7.4

It is water-white, viscous, oil like liquid; density 0.965-0.970; $[n]_D^{25}$ = about 1.404. Viscosity at 25°C about 60,000 cs. Immiscible in water, alcohol, miscible with $CHCl_3$ and ether.

It is the only antifoaming agent permitted for use in edible oils and fats for deep fat frying up to a maximum limit of 10 ppm under Rule 62A as well as the only release agent permitted under Rule 62B for use in confectionary not more than the same 10 ppm. For the purpose of Rule 62A an antifoaming agent means substances which retard deteriorative changes and foaming heights during heating.

Pearson explains this function stating that such chemicals reduce lipid oxidation by preventing oxidative foaming and therefore give the frying oil a longer working life. The antifoaming agent is not soluble in vegetable oils and functions by forming monolayers on the surface of the boiling oil. There are no satisfactory simple methods for the determination of $(CH_3)_2$ polysiloxane in frying oil though Deoden *et. al.* have reported a direct aspiration AAS method with a detection limit of 1 ppm .

Qualitative analysis about the action of antifoam agents in preventing foaming reveals that the agent should be able to displace the foam producing surfactant from the interface. The agent should insert its own poor foam stabilizing characteristics into the system.

While researching on reactions in monomolecular films Mittelmann and Palmer studied oxidation of a double bond in triolein on a dilute $KMnO_4$ substrate and found the area at 8 dyne/cm changed with time as shown in the Fig. 7.4.

The first increase in area (portion I) was interpreted as being due to the reaction.

$$-CH_2-CH=CH-CH_2 \rightarrow -CH_2-CHOH-CHOH-C-$$

with a consequent greater degree of anchoring of the middle of a molecule in the surface region. The rapid decrease in areas that followed, portion II seemed to be due to fragmentation into two relatively soluble species, and the slow residual drop portion III, to some type of polymer formation. The initial action obeyed equation given below and the pseudo first order rate constant was proportional to MnO_4 concentration.

A chemical reaction occurs at constant film pressure where the product(s) are insoluble and remain in monolayer form. Assuming that the area is an additive property, i.e. that the total area will be given by the sum of the areas that would be occupied by the reactant and products separately. Where the change due to the chemical reaction is not much, this assumption may be satisfactory, though like in mixed monolayers serious errors may be involved. Nevertheless on the basis of this assumption.

$$n_A / n°_A = (A - A^{\infty} / Å - A^{\infty}),$$

where "A = No of moles of component A.

If the first step is the rate determining and governed by a constant concentration a first order rate law will be observed. It was found that rate constant fell off with increasing film pressure due to varying probability of the double bond being on the surface.

Actually, reactions in which a product remains in the film (as above) are complicated by the fact that the areas of reactant and product are not additive, i.e. a nonideal mixed film is formed. Gilby and Alexander in some further studies of the oxidation of unsaturated acids found mixed film of unsarurated acid and dihydroxy acid (the immediate oxidation product) were indeed far from ideal[4].

References

1. Franz Timmermann, *International Food Marketing and Technology* 1997, 11(6), pp 19-22.
2. Anderson BMW *FoodAddit. Contam,* 1986,3, pp 225-230.
3. Deoden WG, Kushibale EM and Ingla AC, *JAm Oil Chemists Society,* 1980, pp 57, 73.
4. Adamson W Arthur, *Physical Chemistry of Surfaces,* IIIrd Edition, 1976John Wiley & Sons, INC, pp 162-163.

Chapter 8

Antioxidants

INTRODUCTION

According to Rule 58 of the P.F.A Rules 1955, an antioxidant has been defined as a substance which when added to a food retards or prevents oxidative deterioration of that food and does not include sugar, cereals, oils, flours, herbs and spices.

Safflower, soya and sunflower oils contain a high proportion of unsaturated fatty acids hence are highly prone to oxidative rancidity. Oils of sesame and cotton seed contain in addition to tocopherols, sesamol and gossypol respectively, which are antioxidants therefore are less labile to oxygen attack.

Vegetable oils are subject to spoilage, due to the effect of environmental factors that can affect their stability. These factors are light, atmospheric gases, temperature and moisture.

Of these factors, oxygen is the most critical factor affecting stability. Oxygen may gain access to the oil in several ways. Atmospheric oxygen may be present in the oil. It may also be present in the headspace of the package, or may enter the oil by permeation through the body of seals of the package. The presence of oxygen causes the formation of hydroperoxides which give rise to oxidative rancidity.

Oxidative rancidity occurs in oils having a high degree of unsaturation. Oxidation is catalysed by temperature, very low humidities (moisture from oils), ultra violet light, presence of trace metals (Cu, Fe and Co) and degree of unsaturation. It is also observed that the oxygen susceptibility increases with the distribution of mono-di- and tri-unsaturated fatty acids in the glycerol moiety.

Natural oils and fats are less prone to oxidation due to the presence of antioxidants like tocopherols. Oxidation forms peroxides and hydro peroxides and then aldehydes, ketones and acids which are responsible for off-odour. Rancid oils are nutritionally inferior. The rate of oxidation increases drastically as the number of double bonds increases in the fatty acid chain as well as in the presence of factors like heat and light.

Refining of oils removes the antioxidants and thus refined oils are more prone to rancidity hence antioxidants are added to increase their shelf life.

Although the natural antioxidants found in spices will never compete with their synthetic counter parts, they nevertheless are regarded as beneficial and in some cases significant. Antioxidant activity of some of the spices is, aniseed 1.9; basil 1.2; bay leaf 2.1; small cardamom 1.3; caraway seed 1.8; cassia 1.4; celery seed 1.2; chilli 1.5; cinnamon 1.3; cloves 1.8; coriander 1.3; cumin, dill and fennel seeds 1.3 (each); fenugreek 1.6; ginger 1.8; mace 2.6; mustard 2.0; nutmeg 3.1; paprika 2.5; black pepper 1.4; white pepper 1.2; poppy seed 1.2; thyme 3.0; turmeric 2.9. Rosemary and sage have the highest 17.6 and 16.5 respectively[1].

Under Rule 59, only lecithins, ascorbic acid and tocopherol can be added to any food, however, ethyl, propyl, octyl, dodecyl gallates including gallic acid, citric and tartaric acid, BHA, TBHQ (tertiary butyl hydroquinone, ascorbyl palmitate and resin guaiac are also allowed in very low quantities-0.01 to 0.05%, with certain restrictions.

The remarkable characteristic of antioxidants is their great effectiveness at low concentrations in the oil. The natural antioxidant concentration of an oil is usually about 100 ppm. For example, for the optimum stability of lard at 97°C, the addition of any tocopherol above a concentration of 250 ppm has little or no effect on autooxidative stability reports Parkhurst *et. al*[2].

MECHANISM

No universal mechanism is available to explain how traces of a foreign material are able to exert so powerful a retarding effect on autoxidation. The possible retarding effect of hyoxoquinone and similar compounds including the tocopherols may be as follows where RH = activated linoleic pentadiene system with H as the labile hydrogen of the methylene group, R* = free radical, RO_2* =

oxygenated free radical, RO_2H = primary oxidation product (hydroperoxide),

HO–⟨benzene⟩–OH = hydroquinone

$$HO\text{–}C_6H_4\text{–}OH + R^{\bullet} \rightarrow HO\text{–}C_6H_4\text{–}O^{\bullet} + RH \quad \text{(i)}$$

$$HO\text{–}C_6H_4\text{–}OH + RO_2^{\bullet} \rightarrow HO\text{–}C_6H_4\text{–}O^{\bullet} + RO_2H \quad \text{(ii)}$$

$$\underset{\text{[From (i)]}}{HO\text{–}C_6H_4\text{–}O^{\bullet}} + \underset{\text{[From (ii)]}}{HO\text{–}C_6H_4\text{–}O^{\bullet}} \rightarrow OH\text{–}C_6H_4\text{–}OH + O{=}C_6H_4{=}O \quad \text{(iii)}$$

Figure 8.1

Equations (i) and (ii) show that the antioxidant supplies hydrogen atoms to terminate both the initiating and propagating steps in the antioxidation chain reaction. Because of the energy required to remove H from RH to form R*, this step would be sluggish, and if the antioxidants were present before many radicals and chain carriers were produced, it is easily seen how trace amounts of an antioxidant could greatly retarded the progress of antioxidation even though the antioxidant is eventually destroyed.

A more plausible and detailed mechanisms is: Phenolic (primary) antioxidants, whether naturally occurring, e.g. tocopherols or flavonoids or permitted synthetic compounds, e.g. BHT, BHA, TBHQ and gallates, inhibit chain reactions by acting as hydrogen donors or free radical acceptors. The reaction mechanisms of a primary anti-oxidant, AH are,

AR + RO°	→	ROH + Å	(i)
AH + ROO°		ROOH + A	(ii)
RH + Å	→	AH + R°	(iii)
AH + ROO°		[ROO°AH] Complex	(iv)
[ROO°AH]	→	Non-radical product	(v)
Å + Å	→	AA	(vi)
Å + R°	→	RA	(vii)
Å + ROO°	→	ROOA	(viii)

The inhibitory reaction (ii) is more important than reaction (i) or (iv), and influences the overall inhibition rate content. The stable resonance hybrid of antioxidant free radical A, and the non-radical reaction (v) to (viii) products thus produced are incapable of initiation or propagation of the chain reactions .

Chelating agents and sequestering agents like citric acid and isopropyl citrate, amino acids, phosphoric acid, tartaric acid, ascorbic acid (AA) and ascorbyl palmitate (AP), ethylene diamine tetra acetic acid (EDTA), etc. which chelate metallic ions such as copper and iron, promote lipid oxidation through a catalytic action. The chelators are sometimes referred as synergists since they greatly enhance the action of phenolic antioxidants. It is suggested that the synergist (SH) regenerates the primary anti-oxidant according to the reaction.

$$SH + A \rightarrow AH + S^{\circ} \qquad \text{(ix)}$$

Phospholipids, ascorbic acid and its palmitate are known as antioxidant synergists in oils and food systems. AP and AA also act as oxygent scavengers and the reactions lead to dehydro-products.

$$2AA + O_2 \rightarrow 2\ \text{dehydro - AA} + 2H_2O \quad \text{(x)}$$

$$2AP + O_2 \rightarrow 2\ \text{dehydro - AP-I-}2H_2O \quad \text{(xi)}$$

The chain breaking reactions between AP and hydroperoxides, alkyl or peroxy radicals have also been described.

The use of an antioxidant and the choice of which permitted antioxidant to use, depends upon understanding the reaction mechanisms outlined above. For maximum efficiency, a combination of primary antioxidant, synergist and metal chelating agent is often used. Moreover, the application of chelators and synergists prevents decomposition of hydroperoxides (OH flavour precursors), by their sequestering action on catalytic trace metals, thus retarding the development of rancid off-flavours and odours in oils, fats and lipid containing foods.

If the average chain life in autoxidation is long, the presence of even traces of antioxidants causes a tremendous decrease in the rate of autoxidation. Evidence has been published that semiquinone radicals can be formed and are stable enough to exist for appreciable times. Since the antioxidants are ultimately destroyed during the induction period, there is a rapid increase in the rate of peroxide formation afterwards.

Though hundreds of substances have antioxidant properties but very few are innocuous enough to be considered as food grade materials hence the above mentioned statutory permitted materials and their quantities.

Antioxidants have no effect after the end of induction period; once rapid oxidation has set in, the rate is as rapid in a fat that originally contained antioxidants as in an unprotected fat. Obviously any fat exposed to oxygen will resist rancidification for a shorter or longer time according to the rate at which antioxidants are destroyed under the conditions of storage or test. This is generally the case, though exceptions are there.

On the basis of the chain theory, the protective action of an antioxidant should be in direct proportion to its concentration in the fat or other substrate. This is not so except within limited ranges for a few antioxidants and successive additions of an antioxidant yields steadily diminishing returns, in terms of increasing the stability of the substrate because at higher concentration an appreciable portion of the antioxidant is consumed by side reactions, and thus does not function as a free radical terminator.

Possibility of production of pro-oxidant compounds as a result of decomposition of the antioxidant is there. For example excess concentration of tocopherols than their natural level of occurrence in vegetable oils function as pro-oxidants. Ascorbic, phosphoric, citric and tartaric acid, as well as EDTA have been known for a long time to enhance the effects of natural antioxidants in vegetable oils, whereas they have little effect in animal fats. This is due to their chelating property resulting in de-activation of the metal(s) present in the oil/substrate. To be precise, they eliminate or diminish the deleterious effect of metal-catalysed autoxidation.

1. *Propylgallate*

It is also known as PG, Progallin P, Tenox PG; has crystals of melting point 150°C. Solubility at 25°C in water is 0.35 g/100 ml, in alcohol = 103 g/100 g, in ether = 83 g/100 g, in cotton seed oil at 30°C = 1.23 g/100 g, in lard at 45°C = 1.14 g/100 g. It darkens in the

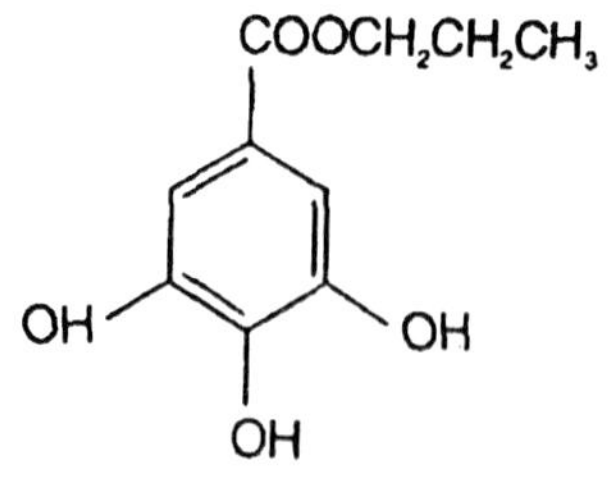

Figure 8.2

presence of iron and its salts. Synergic with acids, BHA, BHT. Antioxidant for foods, fats, oils, ethers, emulsions, waxes, transformer oils.

2. Nordihydroguaiaretic Acid (NDGA)

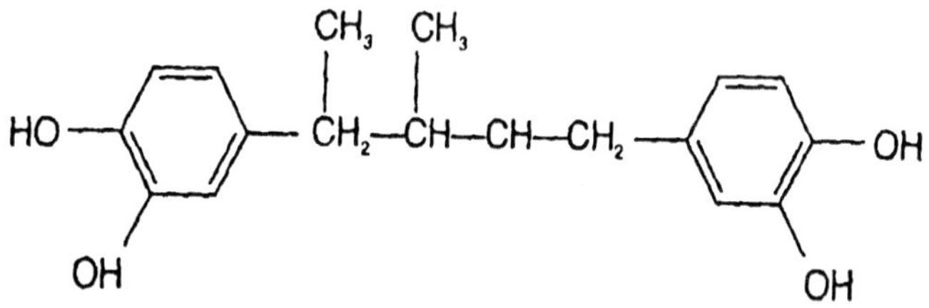

Figure 8.3

It occurs in the resinous exudates of many plants but isolated from the perennial evergreen shrub *Lama divaricata syn. Covillea tridentata.*

It crystallises from dil acetic cid, mt. pt. 184-185°C. Soluble in methanol, ethanol, ether, acetone, glycerol, propylene glycol, slightly soluble in hot water, $CHCl_3$, practically insoluble in pet ether, benzene, toluene, soluble in dil. alkalies, developing deep red colour, insoluble in dil. HCl, soluble in cone. H_2SO_4. Solubility in cotton seed oil at 30° = 7.1 mg/g, in lard at 45°C = 5.2 g/g. Lard containing 0.01% NDGA and stored at room temperature for 19 months in diffused day light showed no appreciable rancidity or colour change.

3. Butylated Hydroxy Anisole (BHA) It is also known as Antrancine 12, Embanox, Nipantiox 1-F, Sustane 1-F, Tenox BHA.

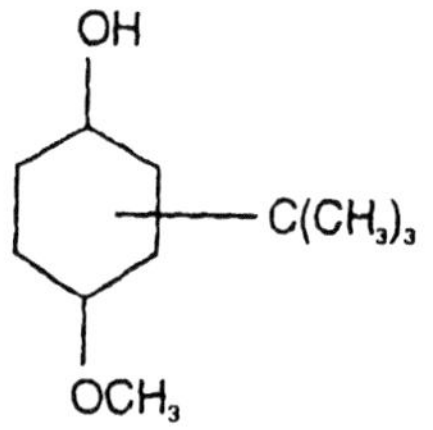

OH

$C(CH_3)_3$

OCH_3

Figure 8.4

It is a waxy solid, mt. pt 48-55°C, insoluble in water, soluble in petroleum ether (Skelly solve H), in 50% alcohol (or higher), in propylene glycol, alcohols, in fats and oils, synergic with acids, BHT, propyl gallate, hydroquinone, methionine, lecithin, thiodi propionic acid, etc. The American- Meat Institute Foundation has proposed an antioxidant mixture known as AIMF-72 which contains 20% BHA, 6% propyl gallate and 4% citric acid in propylene glycol.

4. Butylated Hydroxy Toluene (BHT) It is also known as Antrancine 8, Tenox BHT, Ionol CP, Sustane, Dalpac, Impruvol Vianol.

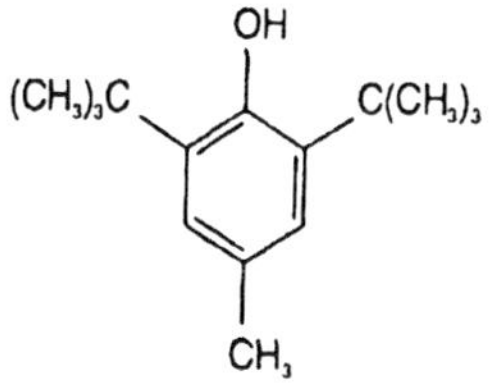

OH

$(CH_3)_3C$ $C(CH_3)_3$

CH_3

Figure 8.5

BHT crystals have mt. pt. 70°C. b.pt. 265°C, $[d]_4^{20} = 1.048$, flash point, (open cup) 127°C; insoluble in water, freely soluble in toluene, soluble in methanol, ethanol, isopropanol, methyl ethyl ketone, acetone, cellosolve, pet ether, benzene, most other hydrocarbon solvents. Solubility in liquid petroleum (white oil) 5% (w/ w/). More soluble in food oils and fats than BHA. Good solubility in linseed oil.

DETECTION OF ANTIOXIDANTS

I. Chemical Method

Reagents

(i) ***Ehrlich reagent*** (Dizazobenzene sulphonic acid 0.5%). 0.5% solution each of $NaNO_2$ in water and sulphonic acid in HCl.

$NaNO_2$ solution is prepared fresh every three weeks. Both the solutions are kept under refrigeration. $NaNO_2$: sulphonic acid :: 1:100 daily.

(ii) ***Dianisidine solution*** 3,3 dimethoxy benzidine. 250 mg is dissolved in 50 ml of anhydrous methanol, 100 mg of activated charcoal is added, mixed and then contents filtered. 40 ml filtrate + 60 ml of 1N HCl are mixed, protected from light, prepared daily and used.

(iii) ***Acetonitrile (CH_3CN) solvent*** It is saturated with pet-ether.

(iv) ***Ba $(OH)_2$*** 1% in water.

(v) ***Activated florisil absorbent*** 80 ± 20 mesh activated at 260 or 650°C. It is tested for BHT retention by adding 0.2 mg of BHT in 25 ml petroleum ether to prepared column, eluted with 150 ml pet. ether. Presence of BHT is tested after evaporating the solvent just to dryness. If BHT is not eluted the remaining florisil is activated by heating for 2 hours at 650°C, cooled, 56.5% water by weight added, homogenised for one hour in a closed container.

The florisil column for clean up of BHT extract is prepared by inserting a small glass wool plug into a chromatographic tube measuring 250 x 20, o.d. mm with stop cock. 12 g of florisil is added by gentle tapping then washed with 2 x 15 ml petroleum ether, adding the second portion when liquid level is just above top of florisil. The column is always kept wet.

Procedure

A. Propylgallate About 30 g of fat/oil is weighed, dissolved in about 60 ml of petroleum ether and transferred to 250 ml separating funnel (SF). 15 ml water is added, contents shaken gently for a minute, layer let to separate, aqueous phase drained into a 125 ml SF, leaving any emulsion in organic phase. The ethereal phase is extracted with 2 x 15 ml of water, the organic phase is further extracted with (iii) above. The aqueous extract is shaken with 15 ml petroleum ether for one minute. The aqueous phase is discarded and the ethereal extract is evaporated just to dryness in small beaker. 4 ml of 50% alcohol is added to residue, swirled then 1 ml NH_4OH is added. A rose colour solution develops which is unstable and fades away after few minutes. However the appearance of rose coloured solution confirms presence of propyl gallate.

B. Nordihydroguaiaretic acid (NDGA) Petroleum ether solution is extracted from (A) by shaking with 20 ml of CH_3CN for two minutes. After layer separation CH_3CN is examined, petroleum ether phase discarded, the combined acetonitrile extract is diluted with 400 ml of water. 2-3 g of NaCl is added, 20 ml petroleum ether is shaken for two minutes, layers left to separate. CH_3CN layer is drained into another one liter SF. The diluted CH_3CN layer is extracted with 2 x 20 ml additional portions of petroleum ether. Dilute CH_3CN solution is kept aside for further extraction. Petroleum ether extracts are combined in a 100 ml beaker and set aside for BHA and BHT tests.

50 ml of pet. ether as well as of ether are added to diluted CH_3CN from (B) and shaken for 2 minutes, layer let to separate, CH_3CN layer discarded and the solvent is evaporated just to dryness in a small beaker. 4 ml of 50% alcohol is added, contents swirled, 1 ml of 1% $Ba(OH)_2$ solution added and remixed. The resultant solution becomes blue and fades rapidly if NDGA is present.

C. Butylated hydroxyanisole BHA 1/3 of combined pet ether solution reserved for BHA & BHT tests is evaporated just to dryness, using gentle heat, under air current. The residue is dissolved in 2.5 ml of alcohol and diluted with 2.5 ml water, swirled and 1ml of (i) is added followed by 1 ml of IN NaOH, swirled again. Formation of a red solution is a positive indication of BHA's presence.

D. Butylated hydroxy Toluene (BHT) The remaining 2/3 from (C) is passed through florisil column and eluted with 150 ml petroleum ether. The eluate is collected in a 200 ml beaker, then evaporated like above to just dryness, 2.5 ml alcohol is added, contents diluted with 5 ml water and mixed. 2 ml of (ii) above is mixed followed by 0.8 ml of 0.3% $NaNO_2$ solution, mixed, kept aside for five minutes then transferred to a small SF. 0.5 ml $CHCl_3$ is added and contents shaken vigorously for half minute, layers let to separate. If $CHCl_3$ layer turns pink to red the BHT is understood to be present subject to confirmation by spectrophotometric curve comparison of coloured $CHCl_3$ extract obtained from reference standard BHT by dissolving approx. 15 mg in 5 ml aqueous alcohol 1:1 and 2 ml of (ii) above.

II. TLC method

The oil sample is dissolved in petroleum ether and extracted with acetonitrile, extract evaporated in vacuum in a rotary evaporator at less than 40°C. The residue is dissolved in alcohol, applied to TLC plates and after development, spots are visualised by spraying with Gibb's reagent.

Reagents

(i) Petroleum ether.

(ii) Acetonitrile.

(iii) Developing solvents.

(a) petroleum ether: benzene : glacial acetic acid :: 2:2:1.

(b) petroleum pether : benzene : ethylacetate : glacial acetic acid :: 40:40:25:4.

(c) $CHCl_3$; CH_3OH : glacial acetic acid :: 90:10:2.

(iv) Spray reagent: 2,6, dichloroquinone chloroimide (Gibb's reagent); 0.1% in alcohol.

(v) Standard solutions (0.1%): Propyl gallate (PG) Octyl Gallate (OG), Dodecyl gallate (DG), BHA and BHT in methanol.

Procedure

10 g of oil/melted fat sample is dissolved in 100 ml of petroleum ether and transferred into a 250 ml SF. 25 ml of acetonitrile is added and contents are shaken gently. Acetonitrile phase is run into another SF followed by three extractions, each time the acetonitrile extract is transferred to a rotary evaporation flask and evaporated at less than 40°C just to dryness, the residue is dissolved in 2 ml methanol.

The developing chamber is saturated with (iii) (a), (b) and *(c)* alternatively. 15 + 5 μl extract solution is applied along with standards (v), (4 μl), on TLC plate coated with silica gel G. The plate is developed to a distance of 15 cms, then left for air drying and finally sprayed with (iv), redried at 103 ± 2°C for 15 minutes. Colour and Rf values are compared with standards. The plate is cooled and in a tank containing ammonia, the characteristic colour changes are carefully observed and compared with the following chart,

Antioxidant	*Rf value with developing solvents*			*Gibb's Reagent*	*Gibb's Reagent followed by Ammonia*
	A	*B*	*C*		
1	*2*	*3*	*4*	*5*	*6*
PG	0.12	0.25	0.55	Brown	Gray-Brown
OG	0.22	0.40	0.64	Brown	Gray-Brown
DG	0.27	0.45	0.66	Brown	Gray-Brown
BHA	0.62	0.87	0.92	Brown-Red	Gray
BHT	0.82	0.99	0.95	Brown-Violet	Gray

Estimation of PG spectrophotometrically Oil or melted fat dissolved in petroleum ether is extracted with amm. acetate solution and water. The combined extract is treated with ferrous tartarate and the absorbance of the coloured solution is read at 540 nm. The amount of PG present in the sample is calculated from the calibration graph.

Reagents

(i) ***Petroleum Ether*** Petroleum ether (PE) 40-60°C : PE 60-80°C :: 1:3 is shaken with 1/10 volume of H_2SO_4, acid layer discarded, ether phase washed several times with water, then once with NH_4OH (1%) solution, again with water until washings are neutral. All washings are also discarded, thus washed petroleum ether is distilled in all glass apparatus.

(ii) ***Ferrous Tartarate*** 100 mg of $FeSO_4.7H_2O$ + 500 mg of $NaKC_4H_4O_6.4H_2O$ (Rochelle salt, i.e. Na-K, tartarate) are dissolved in H_2O and diluted to 100 ml with water. Shelf life only 180 minutes.

(iii) ***Amm. Acetate solutions*** These are 1.25%, 1.67% and 10% aqueous solutions. Solution containing 1.67% amm. acetate in 5% alcohol also prepared.

(iv) ***PG Standard solution*** 50 μg/ml in H_2O.

Preparation of standard curve 7 aliquots of (iv) ranging from 50 to 1000 μg are put in separate marked 50 ml conical flasks. Exact 2.5 ml of 10% am. acetate is added to each flask, diluted exactly to 24 ml with water, 1ml of (ii) is pipetted into each flask. All solutions are kept for 3 minutes, absorbance of each is measured at 540 nm against solution containing 20 ml of 1.25% am. acetate solution + 4 ml H_2O + 1 ml of (ii). From these a calibration curve is prepared*.

Procedure

40 g of oil/melted fat sample is dissolved in (i) and diluted to 250 ml with it. Gentle warming may be done if deemed necessary for complete dissolution. 100 ml of this solution is pipetted into 250 ml SF, extracted with 20 ml of aqueous 1.67% am. acetate solution by gentle shaking for 150 seconds. Layers left for separation, the aqueous layer is drained into 100 ml volumetric flask. Some shortenings are prone to emulsion formation during aqueous extraction. In such cases 2 ml of n-octanol is added to fat solution prior to extraction and 1.67% am. acetate solution in 5% alcohol is used in place of aqueous solution. This procedure is used only when usual method does not work. 2 × 20 ml amm. acetate extractions are collected in a volumetric flask.

Finally the fat solution is extracted with 15 ml water for 30 seconds, the aqueous layer is combined. 2.5 ml of 10% am. acetate solution is added to combined extract and diluted to volume with water, homogenised and filtered through Whatman No. 4 filter paper to remove any turbidity. The colour is developed like before on the same day the extract is prepared. On standing for several hours the combined extracts may develop yellow colour which should be discarded.

Aliquot of extract (20 ml) is pipetted out into 50 ml Erlenmeyer flask, diluted to 20 ml with 1.25% am. acetate solution, 4 ml of water is added, 1 ml (ii) is pipetted. After complete mixing, its absorbance is measured at 540 nm against a solution containing 1.25% am. acetate solution, 4 ml of water and 1 ml (ii).

Amount of PG is calculated from the calibration curve prepared earlier as detailed above.

Note. Excess gallates reduce the intensity of the colour. For example, 200 mg/L of propyl gallate reduces the colour from 200 mg/L BHA to about one half under the standard conditions of the test. The APA report gives a table of corrections to be applied, if necessary.

Spectroscopic Determination of BHA

Reagents

(i) BHA Standard (25 mg/litre in 95% methanol)

(ii) Gibb's Reagent (2,6 dichloroquinone chloroimide), 0.01% in 95% methanol

(iii) Sodium tetraborate decahydrate, 0.5%

Procedure

10 g of warm liquid sample or melted fat + 25 ml of 95% methanol are vigorously shaken for one minute in a centrifuge tube. It is placed in a water bath maintained at 45 ± 5°C for about 15 minutes during which the layers get separated. The upper layer is poured into a 50 ml calibrated flask, another extraction with 20 ml of 95% methanol and transfer of upper layer to the flask is repeated. The contents in the flask are diluted to the mark, 1 g $CaCO_3$ is added, contents mixed by shaking then filtered through Whatmen No. 1 or equivalent filter paper, discarding the first few ml of the filtrate. The amount of $CaCO_3$ is not specific but should permit a clear filtrate. Mix 2 ml of filtrate + 2 ml of 95% methanol + 8 ml of borax solution and 2 ml of Gibb's reagent. After 15 minutes dilute this mixture exactly to 20 ml with n-butanol. Prepare the blank and standard using the 25 mg/ L BHA solution. Read the absorbance at 610 nm and calculate the amount of BHA present in the sample from the absorbance of sample and the standard.

Spectrophotometric determination of BHT

Principle The sample is steam-distilled and BHT in the steam distillate is determined by the colour reaction with O-anisidine and sodium nitrite.

Reagents

(i) ***$MgCl_2$ solution*** 100 g of $MgCl_2$. 6 H_2O in 50 ml H_2O.

(ii) ***O-dianisidine solution*** 250 g is dissolved in 50 ml methanol + 100 g of activated charcoal, together shaken for 5 minutes and filtered. 40 ml of this clear solution + 60 ml of IN HCl, prepared daily is the ready to use reagent protected from light.

(iii) ***Na NO_2 solution*** 0.3% in H_2O.

(iv) ***Standard solution of BHT*** 50 g dissolved in methanol and diluted to 100 ml. Working standards containing 1-5 ug/ml are prepared by diluting with 50% methanol (v/v).

Procedure

15 ml of (i) and 5 g fat in a distillation flask. Oil bath for distillation flask is preheated to 160 ± 10°C. Steam generator is adjusted to distill 4 ml water per minute. Steam generator and condenser are connected to the distillation flask which is immersed immediately in the oil bath. Vigorous steam distillation is commenced, collecting the first 100 ml of the distillate in a 200 ml volumetric flask containing 50 ml methanol. The assembly is dimantled, condenser is washed with 5 ml portions of methanol, adding these washings to the above volumetric flask, cooled to room temperature, volume adjusted to 200 ml with methanol, contents mixed well.

Three 60 ml separating funnels (SF) of low actinic glass or painted black are cleaned and dried. To the first SF is added 25 ml of 50% methanol, to the second SF 25 ml of 1-3 mg/ml standard and to the third one is added 25 ml of distillate. 5 ml of (ii) above is added to each SF, stoppered, contents mixed carefully. 2 ml of (iii) added, stoppered and mixed as before, then left for 10 minutes, then 10 ml of $CHCl_3$ is added to each SF. The coloured complex is extracted by shaking vigorously all the funnels for 30 seconds, layers let to separate. 10 ml volumetric flasks of low actinic glass for blank, $CHCl_3$ layer is drawn off to the corresponding flasks to reach the mark on each flask. With 5 ml $CHC1_3$ + 2 ml methanol as the blank, absorbance of all is read at 520 nm.

$$\text{BHT (ppm)} = \frac{\text{Absorbance of sample} - \text{Absorbanceof blank}}{\text{Absorbance of standard} - \text{Absorbance of blank}} \times$$

$$\text{standard}\,\mu\text{g/ml} \times \frac{200}{\text{wt of sample}}$$

Colourimetric estimation of gallates Gallates are extracted from oils using methanol and are then estimated colourimetrically using amm. Fe^{++} sulphate (AFS) which produces a purplish blue colour with gallates.

Procedure

(a) *Extraction of gallates from oil:* 9-10 g of oil is accurately weighed into a large test-tube and shaken with 25 ml of 95% methanol for 1 minute. The test tube is placed in a water bath at about 45°C for about 15 minutes. The upper layer is poured into a 100 ml volumetric flask, the extraction is repeated with 20 ml of 95% methanol, again transferring the upper layer to the volumetric flask. The volume is made up with water. 1 g $CaCO_3$ is added, contents shaken and filtered, first few ml of filtrate rejected. If the filtrate is not clear it is repeated with more $CaCO_3$.

(b) *Estimation of gallates:* Gallate solutions of 0, 10, 20, 30, 40 and 50 ppm are prepared by dilution of the standard stock gallate solution containing 500 ppm n-propyl gallate. To exactly 20.00 ml of each of these solutions, and to exactly 20.00 ml of the oil extract, 2 ml of acetone and about 10 mg of finely powdered AFS is added, contents shaken for a minute then set aside for 15 minutes. The absorbance of each solution is measured on a spectrophotometer at 580 nm or on a colourimeter using a red filter.

Calculation

A calibration graph is prepared of absorbance against ppm gallate, then concentration of gallate in the original oil is determined as,

ppm gallate in oil = 100 A/W, where

A = concentration of extract in ppm and

W = weight of oil extracted.

References

1. Parrel Kenneth T, *Spices, Condiments and Seasonings* (An AVI Book), 1990, pp 238, Van Nostrand Reinhold. New York.
2. Parkhrust RM, WA Skinner, and PA Sturm, *J. Am. oil chemists soc* 1968, pp. 45, 641-642.
3. Saxby MJ (Ed.), *Food Taints and Off-Flavours,* 2nd edition, Blackie Academic and Professional, London, 1996.

Chapter 9

Vitamins and Minerals in Foods

VITAMINS IN FOODS

Vitamin A Present in animal foods as retinol, and in plant foods mostly as all *trans* β-carotenes. Their chemical structure includes double bonds which are susceptible to oxidation. Attacked by peroxides and free radicals formed from lipid oxidation. Losses promoted by traces of copper and iron which catalyse the oxidation. Negligible losses due to leaching. Heat converts part of the *trans* isomer to neo-β carotene μ which has lower potency.

Vitamin B_1 (Thiamin) Present in animal and plant tissues either as free thiamin or bound to pyrophosphate or protein. Destroyed by sulphur dioxide (in sulphited fruits and vegetables), $KBrO_3$ flour improver and thiaminase. Polyphenoloxidase catalyses thiamin destruction by phenols in plant tissues. Substantial losses due to leaching and drip losses.

Vitamin B_2 (Riboflavin) Occurs as free form in milk but mostly bound with phosphate in other foods. Destroyed by alkaline conditions, light and excessive heat. Stable in air and acids.

Niacin Occurs as nicotinamide adenine dinucleotide and nicotinamide adenine dinucleotide phosphate and as nicotinic acid. Bound to polysaccharides and peptides hence not available in many cereals unless liberated by heat or alkaline conditions like baking powder. The amino acid tryptophan is converted to niacin in the body (niacin equivalent is free niacin + 1/6 of the tryptophan). Generally stable.

Folic acid Occurs in various forms, expressed as pteroyl glutamic equivalent, with various numbers of glutamate residues and methyl or formyl groupings. Richest sources are dark green leaves, liver and kidney. More difficult to assay than other vitamins. Possibly one of the few causes of deficiency disease in industrialised countries, especially in pregnant women, preterm infants and the elderly.

Vitamin B_6 (Pyridoxine) Occurs in three forms, namely pyridoxine, pyridoxal and pyridoxamine. The first two are found in plants and the last two in animal tissues. Mostly in a free form in milk, otherwise bound. Difficult to assay and may be deficient in some diets. Lost by reaction with sulphydryl groups of proteins and amino acids when heated or during storage.

Vitamin B_{12} (Cyano cobalamin) Small losses due to interaction with vitamin C and sulphydryl compounds in the presence of oxygen in milk. Generally stable.

Vitamin C (ascorbic acid) Occurs as both ascorbic acid and dehydroascorbic acid. The latter is very heat labile, with or without the presence of oxygen. Very soluble and readily lost by leaching and in drip losses. Destroyed by a number of plant enzymes, including ascorbic acid oxidase, peroxidase, cytochrome oxidase and phenolase. Copper and iron catalyse oxidation in air, but SO_2 protects against oxidation. Most labile of the vitamins and substantial losses in most processing. Vitamin C retention sometimes used as an indicator of severity of processing. Also used as an antioxidant and stabilizer, as a flour improver and in cured meats.

Vitamin D Occurs in foods as cholecalciferol (D_3) and produced in the skin under the influence of UV light. The synthetic type, ergocalciferol (D_2) is added to save milk products, baby foods and margarine. The vitamin is stable under all normal processing and storage conditions.

Vitamin E Occurs in eight compounds: 4 tocopherols and 4 tocotrienols, each of which has a different potency. Activity usually expressed as α-tocopherol equivalents. Naturally occurring antioxidant but is lost very slowly. Generally stable during processing, except frying in which it is destroyed by peroxides.

Stability of Vitamins in Food

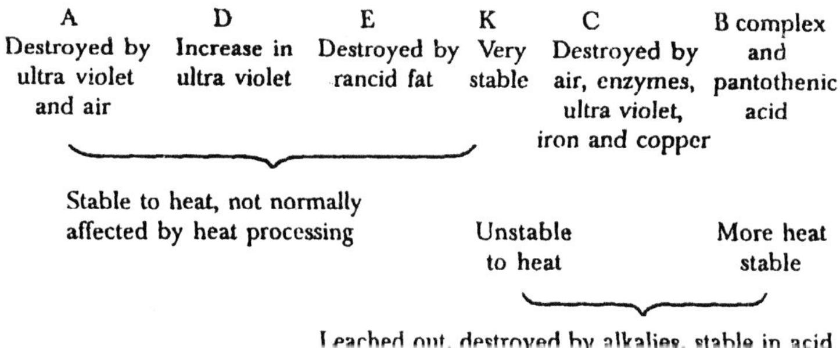

Estimation The estimation of individual vitamins in foods is, generally speaking, more difficult than that of most other food constituents. This is due to the often exceedingly small amounts present in food and to the diverse and complex nature of most vitamins.

Methods available for vitamin assay include the following,

(i) ***Microbiological*** Many of the β-complex vitamins, such as thiamin and riboflavin, may be estimated by measuring the growth of microorganisms having a specific requirement for the particular vitamin being assayed.

The procedure involves the preparation of a calibration curve from a set of standards of known vitamin in concentrations and estimating the vitamin content of the food extract by comparison against these standards. The growth of the microorganisms may be measured on the basis of techniques such as CO_2 production or turbidity measurements.

(ii) ***Fluoritmetery*** Some vitamins, such as riboflavin, are naturally fluorescent and may be estimated by measuring the degree of fluorescence of a food extract and comparing against a set of standards. Others may be converted to fluorescent derivatives and the same process under taken, e.g. thiamine may be converted to the fluorescent thiochrome.

(iii) ***Colourimetery*** Vitamin A, known as retinol may be estimated colourimetrically by reacting a food extract with $SbCl_3$ which converts the vitamin into a coloured product, whose absorbance may be measured and compared against standards.

(iv) ***Titration*** Vitamin C, known as ascorbic acid, may be titrated against the dye DPIP (2, 6-dichlorophenol indophenol) which

is reduced by Vitamin C from a blue colour to a colourless reduced form. By standardising the dye against a solution of Vitamin C of known concentration, food samples may then be titrated and analysed. This is a simple, rapid routine method suitable for analysing fruit juices, wines, etc.

(v) ***HPLC (High Performance Liquid Chromatography)*** Many vitamins may now-a-days be routinely estimated by the use of HPLC, the choice of column material, detector, solvents and instrumental parameters being geared to the particular vitamin concerned. This procedure has provided a greater degree of sensitivity, precision and accuracy than was otherwise available with most of the alternative methods outlined above.

Estimation of Vitamins in foodstuffs Vitamins are required to be assessed in a large number of foodstuffs like dairy products, processed cereals and many others also. Different methods are used for vitamin assay in different laboratories which prevented a comparative study of results. Bureau of Indian Standards in consultation with the user industries, consumer interests, enforcement agencies of the government about food safety and quality has brought out uniform assay methods for vitamins and these have been summarised. These methods are mainly based on practical experience by the above groups within India.

ESTIMATION OF CAROTENES AND VITAMIN A (RETINOL)

Preparation of assay sample The sample should be representative of the whole and any deterioration of the vitamin ,to be examined should be prevented. Powders and liquids should be thoroughly homogenised and dry materials are pulverised and mixed thoroughly. Butter is melted under constant stirring. Samples from margarine/ cheese/other such foods should contain portions of the surface as well as of the interior. Wet / fresh materials may be minced with a knife or scissors, or homogenised in a blender, if essential, in the presence of the extracting solvent.

Estimation of carotenes Carotenes are extracted from the moist pulverised or homogenized assay sample with pure hexane, petroleum ether or other suitable solvents mixed with acetone or alcohol.

Fatty materials must be preliminary saponified. The extract is purified from non-carotene pigments chromatographically. The carotene content of aliquot from the elute in hexane or petroleum ether is estimated spectrophotometrically at 450 mu or in a photoelectric colourimeter with a 440 mu filter.

Extraction vessel Depending on the nature of the assay sample, a glass mortar for moist homogenised food material, a beaker with glass rod with a flat end for maceration and Bailey-Walker type or similar hot extraction apparatus for dried leaves and powdered beans is selected. The chromatographic tube should be 200 ± 20 mm long having a diameter of 20 mm.

Reagents

(i) *Petroleum ether (PE): acetone mixture*

(a) For extraction 1:1 (v/v) and (b) for elution 9:1 (v/v).

(ii) *n-Hexane/petroleum ether (PE)* It is purified by shaking with cone. H_2SO_4 and the fraction distilling between 67 to 70°C is collected.

(iii) *Diethyl ether* Washed with saturated. $FeSO_4$ and then distilled over NaOH.

(iv) *Granular anhydrous Na_2SO_4* conforming to 18:255-1967.

(v) *Adsorbent* Activated magnesia: Hyflosupercel or equivalent:: 1:1 is kept ready in dried state till used.

Procedure

(a) Extraction A quantity containing 0.05 to 0.10 mg of carotene is accurately weighed and transferred to the extraction vessel detailed earlier. An equal volume of quartz powder or acid washed sand is added, the mixture is moistened with a little water. 10 ml of (i) is added, contents are grinded, the solvent is decanted into a separating funnel containing water. Normally 4 to 5 such extractions are adequate. The extract is washed with 5 x 100 ml portions of water to remove acetone completely lest it may interfere with chromatography. The extract, containing carotenoids and chlorophylls is preserved for chromatography.

Carotenes are extractable much more conveniently from wet materials. During extraction the deficiency of moisture (taken away by acetone) is made up by adding a few drops of water between the extractions. The cold extraction process does not extract carotenes easily from dry leafy materials or beans hence hot extraction is imperative. With fatty materials, when fat: carotene :: 1000:1, the fat is hydrolysed like in Vitamin A; the non-saponifiable matter is taken in PE or hexane.

(b) ***Purification*** The chromatographic tube is connected to a Witz's or equivalent filtering apparatus with a 150 ml beaker as receiver or to a suction flask of 250 ml capacity, through a rubber stopper. The lower end of the chromatographic tube is plugged loosely with glass wool or fat free cotton and suction is turned on. Reagent (v) mentioned earlier is added through a funnel in small amounts to a height of 11 ± 1 cms, the column is packed by pressing down with a cork stopper, just fitting the tube and attached to a glass rod. 2-3 cm layer of reagent (iv) is placed on the top.

Chromatography PE/hexane extract is poured slowly into the absorption tube, the vessel is carefully rinsed with PE so that all colouring matter is transferred to the column. Lastly, the down sides of the tube are rinsed with a few ml portion of PE/hexane, taking particular care that the chromatographic column containing carotenes always remain covered with the solvent. When a few ml of PE/hexane has come through the column, the receiver of the suction flask is changed. As soon as PE/hexane layer in the column has come down to the Na_2SO_4 layer, 50 ml of reagent (i) (b) is slowly added to elute the carotene. After this mixed solvent has passed through the Na_2SO_4 layer, suction is withdrawn, the elute is transferred to the 10/25 ml volumetric flask as required and diluted to volume with PE.

Preparation of Standard curve The standard curve is necessary only when the final estimation is carried out in a colourimeter or when the true optical density cannot be obtained from the spectrophotometric readings.

About 15 ± 5 mg of purified crystals of β-carotene are weighed accurately and dissolved in about 20 ml of diethyl ether. These are transferred into a 1 litre volumetric flask with about 20 ml of diethyl ether, the volume made up to the mark with n-hexane/PE. From this stock solution, eight concentrations of β-carotene containing 0.25,0.50,0.75, 1.00, 1.25, 1.50,1.75 and 2.00 mg/litre are examined.

Absorbance of these solutions is found out in the instrument with the same cuvettes, to be used for final estimation, at 450 mμ in case of spectrophotometer or with 440 mμ narrow band filter in the case of colourimeters. Absorbance v/s concentration of carotene/ litre is plotted. This relationship should be almost linear.

Determination of Carotenes The absorbance of the hexane/ PE extract of the assay sample is determined like under chromatography as above. From the observed absorbance, the concentration of carotenes in mg/litre is calculated. In case the absorbance is more than 0.50, suitable dilution is made to get the absorbance below 0.50.

Calculation

Content of carotene (mg)/100 g of assay sample A

(i) **From standard curve:** A = CV/10 g, where C = concentration of carotene in mgm/litre from the standard curve, V = final volume of extract, and g = weight of the sample.

(ii) **From optical density (E) at 450 mu**: A = 100 E/196 LW, where L = length of light path in the cuvette in cm, and W = weight in g of the sample/ml of final extract.

Note 1: In case carotene has to be calculated directly without the use of standard curve, the spectrophotometer should be calibrated for wave length and absorption by standard procedure.

Note 2: The factor 196 is for average mixture of carotenes in food samples:

$E^{1\%}_{1cm}$ = 196° in petroleum ether for mixtures of carotenes commonly present in food stuffs and for pure all trans β-carotenes.

$E^{1\%}_{1cm}$ = 2550, in petroleum ether.

Note 3: When the carotene mixture is further purified by chromatography and all trans β-carotenes are isolated and estimated, the following formula should be used.

A = 100 E/225 LW.

Estimation of Vitamin A (Retinol)

Spectrophotometric method Vitamin A is extracted from the assay sample, purified by saponification or chromatography and dissolved in optically pure isopropanol, hexane or cyclohexane. This method is based on the measurement of light absorption of the vitamin solution. This light absorption is proportional to the concentration of the vitamin at a wave length of 325 mμ where maximum absorption occurs. To correct for extraneous materials which also absorb in this region, readings are also made at 310 mμ and 334 mμ followed by appropriate mathematical correction, in case of samples having an absorption maxima for the unsaponifiable matter lying within the region 323 mu and 327 mu and extinction ratio $E_{300}^{mu} / E_{325}^{mu}$, not exceeding 0.73. But for those samples having an absorption maxima for the unsaponifiable matter lying outside the region 323 mu to 327 mu and extinction ratio $E_{300}^{mu} / E_{325}^{mu}$, exceeding 0.73, further purification by chromatography is necessary.

The chromatographic tube has three sections fused together at the upper part, it has wider section, 80 mm length and 15 mm i.d. with a side tube attached at 45 to 50 mm below the top; this side tube has 3 way cock. The central part of the tube is 180 to 200 mm long and has 10 mm i.d., at the lower end it has a stem, 50 mm long and 5 mm dia. The side tube at the top is connected through a rubber tubing to a pressure system consisting of a Büchner flask and a rubber hand-bellows with a non-return valve, so that pressure inside the tube may be applied or released as and when required. A 50 ml separating funnel (SF) is fitted at the top with a wooden cork (not rubber) to serve as a reservoir for solvents.

An UV lamp for detection of vitamin A zone. It is mercury vapour lamp with a light filter to transmit 320 to 340 mu.

Regents

(*a*) ***Ethanol*** It is purified by refluxing absolute ethanol for 30 minutes in presence of 10 g KOH and 10 g aluminium powder/ litre of ethanol. The mixture is then distilled in an all-glass apparatus with a fractionating column of 55 ± 5 cm length. Only middle section of the distillate is collected, the first and the last fractions are discarded, each amounting to 5% of the total volume. The ethanol thus obtained should have the same optical absorption as glass distilled water in UV region.

(*b*) ***Isopropanol*** It is purified by redistillation in an all-glass apparatus with a fractionating column, of 40 cm length and collecting the fraction distilling at 82.3 ± 0.2°C.

(c) ***n-Hexabene/PE*** It is purified like at (ii) under Retinol.

(d) ***KOH - 50% (w/v)***

(e) ***Ether*** It is peroxide free and redistilled. Freedom from peroxides is achieved as described later on.

(f) ***Na_2SO_4*** It is anhydrous and granular. It does not absorb vitamin A under experimental conditions, and 10% solution is not acidic to methyl red indicator solution.

(g) ***Aluminium Oxide*** Neutral aluminium oxide, passing through 100 ± 25 micron sieve, is heated for chromatography at 500 ± 50°C for 2 to 3 hours and protected from moisture by cooling it under vacuum and storing it in an airtight bottle. 50 g of this dehydrated alumina is shaken very thoroughly with 2.0 to 2.5 ml of water till a uniform lump-free mass is obtained. It is kept for 2.5 ± 0.5 hours before use in chromatography.

Solvents for chromatography

(i) ***Petroleum ether*** Its boiling point should be between 67 to 70°C and purified as done for retinol estimation.

(ii) ***Elution mixture*** It is petroleum ether containing 20 ± 4% (v/ v) diethyl ether. The amount of diethyl ether necessary, depends on the activity of alumina, this is found out, if needed, by occasionally checking the vitamin A zone with the weak UV lamp described earlier. The vitamin A zone shall reveal itself by greenish yellow fluorescence.

Procedure

Adequate precautions are essential to protect vitamin A content in the flask from direct actinic light which causes photo chemical changes. Not more than 7 g of the test material is accurately weighed so that it should have 40 ± 10 I.U. of vitamin A. The weighed material is quantitatively transferred to an all glass saponification flask 150 ml, if necessary with a little alcohol. 30 ml of ethanol [95% (v/v)] + 5 ml of (iv) above. A stream of nitrogen gas is passed through the solution to replace the air with jet tube for 5 minutes. The contents should be saponified in a dark corner under a reflux in an atmosphere of nitrogen, over a water-bath, electrically heated, for 35 ± 5 minutes. The contents are cooled, 30 ml water added. The contents are quantitatively transferred to the SF (conical squible shaped, 250 ml capacity). The flask is washed twice with 10 ml water, the washings also added to the SF. The saponified solution is extracted thrice with 50,20 and 20 ml of ether, ether extracts combined in another SF. There contents are washed with 50 ml ice cold water containing a little $NaHSO_4/KH_2PO_4$, just enough to neutralize the alkalinity of the extract. The washings are repeated with ice cold water only, till the washings are neutral to phenolphthalein.

The above washed ether extract is transferred to a stoppered 100 ml measuring cylinder, the SF rinsed with a little ether, the rinsings also added to the cylinder. The volume is made up to the mark with ether, 15 g of (f) added, contents shaken mildly then let to settle in a dark cool place.

Chromatography An aliquot of the ethereal extract of the unsaponifiable matter is evaporated in a nitrogen stream. The residue is taken up in 5 ml of PE, a wad of fat-free cotton is placed in the lower stem of the chromatographic tube then set up in a stand. PE is poured in upto the middle of the 10 mm section, followed by sufficient (g) with continued tapping, to a height of 100 mm in the column and (f) to make a 1 cm layer. SF is attached at the top and pressure applied to drain off excess solvent with particular caution that at no time the level of solvent falls below the level of (f) in the column. Pressure is released, SF removed, the PE extract of the unsaponifiable matter transferred quantitatively to the column, using more PE if needed. SF is replaced, 30 ml PE added gradually in 3 x 10 ml portions through the SF, pressure regulated to obtain a flow rate of approximately 60 drops/minute. This washing would contain carotene fraction, hence collected and estimated, if necessary. The receiver is replaced, 30 ml elution mixture of appropriate composition as detailed therein, is added, eluted carefully and preserved for estimation.

The ethereal extract of the unsaponifiable residue is evaporated or eluted from the chromatographic procedure, using moderate heat and in a dark place away from actinic light in a stream of nitrogen. The residue is taken up in sufficient (b)/(c) to give the concentration, expected to yield absorbance reading of 0.2 to 0.5 at 325 mμ. Absorbance (E) of this solution is determined at 310, 325 and 334 mm.

Calculation

(i) E_{325} (corrected) = 6.815 E_{325} - 2.555 E_{310} - 4.260E_{334}

(ii) Vitamin A content in I.U/100 g of the sample

$$\frac{E_{325} \text{ corrected} \times 183000}{LC}$$; where,

L = length of light path in absorption cell in 1 cm, and
C = amount of assay sample (g/100 ml) of isopropanol.

Note: This formula is applicable only when isopropanol/hexane is used. In case cyclohexane is used, then the factor 1737 should be used instead of 1830.

Carr-Price method The assay sample is saponified with ethanolic KOH and Vitamin A extracted with PE. After reaction with $SbCl_3$, the blue coloured formed is measured in a colourimeter at 610 to 620 mμ. Calculation of Vitamin A content is carried out with reference to a calibration curve.

Special Reagents

(i) ***Ether (peroxidefree)*** may be maintained free from peroxides by adding wet zinc foil, approximately 80 cm^3/litre, cut in strips long enough to reach at least half way up the container; the zinc strips should have been previously immersed in-diluted acidified $CuSO_4$ solution for one minute and subsequently washed with water.

(ii) ***Chloroform*** For use in this determination, it is by washing it thrice with fresh 10% aqueous solution of $Na_2S_3O_3$ in a SF, drying it with anhydrous $CaCl_2$ then filtered and distilled over anhydrous $Na_{,,2}S_2O_3$ in an all glass apparatus with a fractionating column of 70 cms length. Only middle fraction of the distillate is collected, the first 10% as well as the last 10% of the distillate is discarded.

Anhydrous $Na_2S_2O_3$ is prepared from crystalline material by heating it between 105 to 110°C and stored in a desiccator.

(iii) ***$SbCl_3$ Solution*** It is prepared by dissolving 113.4 g $SbCl_3$ into 350 ± 50 ml $CHCl_3$, (ii) above. 5 g of anhydrous $CaCl_2$ is added, contents heated and filtered while hot. The filtrate is diluted to 500 ml with $CHCl_3$.

Procedure

Adequate protection for vitamin A content in the flask from direct actinic light is must.

(i) ***Saponification*** Pure Vitamin A reference standard capsules 0.5 to 1.0 g is accurately weighed. This is transferred quantitatively to an all glass saponification flask 150 ml, if necessary with a little alcohol. 30 ml of ethanol (95% v/v), and 5 ml of KOH 50% (w/v) is added. A stream of nitrogen gas is passed through the above solution with a jet tube for 5 minutes to replace air. In a dark corner, the contents are saponified under reflux over an electrically heated water-bath for 35 ± 5 minutes. After saponification the contents are cooled, 30 ml H_2O added and transferred quantitatively to a conical, squible shaped, 250 ml capacity separating funnel (SF). The sap. flask is washed twice with 10 ml H_2O, the washings also added to the SF. The saponified solution is extracted thrice with 50, 20 and 20 ml of ether respectively. These ether extracts are combined in another separating funnel and washed with 50 ml of ice cold water containing a little $NaHSO_4$ or KH_2PO_4 just enough to neutralize the alkalinity of the extract. The washings are repeated with only ice cold water, till the washings are neutral to litmus.

The ethereal extract is transferred to a stoppered measuring cylinder (100 ml), SF is rinsed with a little ether, this rinsing also added to this cylinder. The volume is made up to 100 ml with more ether. Anhydrous Na_2SO_4 granular in form and not capable of absorbing vitamin A under conditions of experiment, whose 10% solution shall not be acidic to methyl red indicator solution, 15 g is added, contents shaken mildly for a little while, then contents are left to settle in a dark cool place.

(ii) ***Preparation of the calibration curve*** *A* suitable aliquot of the ether solution of the unsaponifiable extract is evaporated to about 5 ml. The remaining ether is evaporated off at low heat under reduced pressure. The residue is taken up in sufficient $CHCl_3$ so that after addition of $SbCl_3$ solution an absorbance of about 0.8 in the photoelectric colourimeter is obtained. From this stock solution of the standard, a series of dilutions in $CHCl_3$, are made to give absorbance values of 80, 60, 40 and 20 percent of the original absorbance. Absorbances of the blue colour formed when 1ml aliquot of each of these five solutions plus 1 ml of $CHCl_3$ with a drop of acetic anhydride is treated with the volume of $SbCl_3$ solution, that is, suitable for the operation and hereinafter referred to as the *faced volume.* The blank is adjusted to 100% transmittance using a tube containing 2 ml of $CHCl_3$, and the *fixed volume* of the $SbCl_3$ solution.

Using a rectangular coordinate paper, the five absorbances obtained against known quantities of vitamin A are plotted, the best smooth curve from the origin through these points is drawn, with no attempt to draw straight line unless the curve is in fact a straight line with the origin at zero. Further those instruments that provide other than straight line curve, this curve is checked at frequent intervals, those who do provide straight line calibration curve, readings of the reference solution are made with each set of sample readings to establish the curve. In the latter case, the calibration curve is re-established whenever variation in the reagent or other variables occurs in procedure.

Determination Not more than 5 g of the material is accurately weighed containing 20 to 45 I.U. of vitamin A, saponified as above, residue obtained after evaporating the ether under moderate heat and reduced pressure. The residue is dissolved in a definite volume of $CHCl_3$ so that 2 ml of the $CHCl_3$ solution with *the fixed volume* of the $SbCl_3$ solution would give an absorbance of about 0.5 to 0.2. The instrument is set at 100% transmittance with 2 ml of $CHCl_3$ + *fixed volume* of the $SbCl_3$ solution as blank. The tube containing 2 ml of $CHCl_3$ solution of the residue is placed, a drop of acetic anhydride is added followed by rapid addition of $SbCl_3$ solution with the help of vacqupet or a similar device. The maximum colourimetric reading is recorded. Vitamin A content is determined from the standard curve and units of vitamin A/100 g of the sample are calculated.

Note: Care must be taken to note the readings within 15 seconds after the addition of $SbCl_3$, solution.

Detection and estimation of carotenoids

1. ***HPLC method*** HPLC is the method of choice for both qualitative and quantitative work. The usual HPLC systems employed for the purpose are detailed here.

(a) ***Normal phase (Adsorption) HPLC*** Gradient elution is usually employed for extracts that contain carotenoids of widely different polarities. Impressive resolution of carotenediols and more polar xanthopylls can be achieved with the following programme

(i) 0-8 minutes isocratic, solvent A : solvent B :: 10 : 90.

(ii) 8-20 minutes linear gradient, solvent A 10-25%.

(iii) 20-30 minutes isocratic, solvent A 40% and

(iv) 30-40 minutes isocratic, solvent A 70%, where,

solvent A is hexane: propan 2-ol (8:2) and solvent B is hexane and thé flow rate is 2 ml/minute.

This procedure will separate efficiently mixtures of geometrical isomers of the main xanthophylls such as zeaxanthin, lutein, viola-xanthin and neoxanthin. The use of propan 2-ol is not so satisfactory, for separating less polar carotenoids. However, for this, mixtures of hexane with ethyl acetate are recommended, but long pre-equilibration is needed to remove residual traces of more polar solvents, especially alcohols. Bonded nitrile phase columns (e.g. spherisorb S5-CN) permit the very efficient resolution of mixtures of similar carotenoids, including structural isomers (lutein and zeaxanthin), geometrical isomers, 5,6- and 5,8-epoxides and even the 8 R and 8S epimers of 5,8 epoxides. The best results are achieved by isocratic separation of fractions (e.g. dihydroxyfraction) that have been obtained from natural extracts by classical column chromatography or TLC. Combinations of two solvent mixtures are used; solvent A is of constant composition, namely hexane containing 0.1 % N-ethyl-di-isopropylamine, whereas solvent B comprises dichloromethane containing a variable proportion of methanol. Solvent composition recommended for separating the carotenoids of various polarity groups have been tabulated by Riiedi, P, *Pure Applied Chemistry,* 1985, pp 57, 793-800.

(b) ***Reversed-phase HPLC:*** Reversed phase partition chromatogra-phy is now most widely used for the routine analysis of carotenoids in natural extracts'.

There is virtually no risk of decomposition or structural modification even of the most unstable carotenoids. The stationary phases usually used are there with C_{18}-bonded chains and the resolution is somewhat dependent on the extent of end-capping and carbon loading of the stationary phase.

The order of elution from reversed-phase columns is not the exact opposite of that of normal-phase columns. For instance, lutein is elüted before zeaxanthin in both methods. The overall polarity of compounds, i.e. presence of polar substituent groups and other important factors influence separation. The strongest association with the C_{18} stationary phase is that of an unsubstituted carotenoid end group, so that a diol with both hydroxy groups in one end group will be substantially more strongly retained than one with one hydroxyl in each end group.

Acyclic carotenoids generally have longer retention times than cyclic ones containing the same functional groups, and carotenoids with a greater degree of saturation are usually more strongly retained.

Non-aqueous solvent systems, e.g. acetonitrile - CH_2Cl_2-CH_3OH have been widely explored for reversed-phase HPLC of carotenoids as reported[2]. G. Britton has employed a linear gradient of ethyl acetate (0-100)% in acetonitrile-H_2O (9:1, containing 0.1% $(C_2H_5)_3N$) at a flow rate of 1ml/minute over 25 minutes, and has obtained good resolution both of polar xanthophylls and of non-polar carotenes, carotene epoxides and xanthophyll acyl esters in the same run. The gradient is easily modified, if needed, to improve resolution of selected parts of the chromatogram[3]. A naturally esterified carotenoid will usually contain a mixture of molecular species containing different esterifying fatty acids. The presence of esters is, therefore, usually revealed by a pattern of usual peaks in the chromatogram in the vicinity of β-carotene. The absorption spectra of xanthophylls are not altered by esterification so the carotenoid components can tentatively be identified.

(c) ***HPLC of Apocarotenoids*** The chromatographic behaviour of apocarotenoids on a reversed-phase column can cause some confusion. The retention times are short; apo-p carotenals, for example, are found in the same area of the chromatogram as the very polar xanthophyll, neoxanthin.

The retention time increases with increasing chain length, which is a more important determining feature than the nature of the functional group present.

2. UV Visible Light Absorption Spectroscopy (UVLAS) This spectroscopy is the basic spectronic method used as a first identification criterion for carotenoids because the position of the absorption maxima, λ_{max} and the shape or fine structure of the spectrum are characteristic of the chromophore of the carotenoid molecule. It is pertinent to remember that the spectrum provides information *only about the light-absorbing chromophore and not about other structural features of the molecule, such as the presence of functional groups.* For instance, OH-carotenoids generally have spectra identical to that of the parent hydrocarbon.

Position of the Absorption Maxima (λ_{max}) The absorption spectra of most carotenoids exhibit three maxima. Values of λ_{max} are markedly dependent on solvent. The values recorded in petrol, hexane and ethanol are almost identical, but values recorded in acetone are greater by around 4 nm, in $CHC1_3$ or benzene greater by 11 ± 1 nm, in CS_2 greater by 37 ± 3 nm. When spectra are determined on-line during HPLC, it must be remembered that the values for λ_{max} in the eluting solvent frequently do not correspond to the published values recorded in a pure solvent. Besides, with gradient HPLC, the solvent composition is changing through out the chromatography, so compounds which in published tables are given the same maximum values may not give the same maximum values during chromatography.

In any given solvent, λ_{max} values increase as the length of the chromophore increases. Non-conjugated double bonds like C-4, 5 double bond of the 6-ring, do not contribute to the chromophore. Extension of the conjugated double bond system into a ring (the C-5, 6 double bond of the p-ring) does extend the chromophore but, as the ring double bond is not coplanar with the main polyene chain, the λ_{max} occur at shorter wave lengths than those of the O acyclic carotenoid with the same number of conjugated double bonds. Despite their being conjugated undecaenes, the acyclic, monocyclic and dicyclic compounds lycopene, γ-carotene and β-carotene have λ_{max} at 444,470,502 nm, at 437,462,494 nm and at 425, 450, 478 nm, respectively.

Carbonyl groups, in conjugation with the polyene system, also extend the chromophore. All such substituted carotenoids therefore have λ_{max} virtually identical to those of the parent hydrocarbon with the same chromophore. For example β-carotene, and its OH-derivations β-crypoxandiin, zeaxanthin, isocryptoxanthin and isozeaxanthin, all have virtually identical spectre with λ_{max}, at 425, 450, 478 nm.

Spectral fine Structure The overall shape or fine structure of the spectrum (equivalent to persistence) is also diagnostic and generally reflects the degree of planarity that the chromophore can achieve. Thus the spectra of acyclic compounds are characterised by sharp maxima and minima. The degree of fine structure decreases a little when the chromophore exceeds nine double bonds. Cyclic carotenoids in which conjugation does not extend into the rings have simple linear polyene chromophores.

When conjugation extends into a β-ring, steric strain causes the ring to adopt a conformation in which the ring double bond is not coplanar with the π-electron system of the polyene chain. This effect is even more pronounced in carotenoids that have carbonyl groups in conjugation with the polyene chain. Spectral fine structure therefore decreases in the order lycopene > γ-carotene > β carotene > canthaxanthin. Canthaxanthin shows only a single, rounded, almost symmetrical absorption peak (in ethanol), but a slight degree of fine structure remains if the spectrum is determined in a non-polar solvent such as light petroleum.

Geometrical homers For Z-isomers, the λ_{max} are generally 1-5 nm lower, the spectral fine structure is decreased and a new absorption peak, usually referred to as the "Cis-peak" appears at a chracteristic wave length in the UV region 142 ± 2 nm below the longest wave length peak in the main visible absorption region. The intensity of the *Cis-peak is* greatest when the Z-double bond is located at or near the centre of the chromophore.

Quantitative Determination

(i) ***Spectrophotometery*** This analysis is normally used for the quantitative determination of carotenoids. The amount of carotenoid present is calculated from the equation:

$$x = Ay \,/\, (A^{1\%}_{1cm} \times 100) \text{ where,}$$

x = mass of the carotenoid (g),
y = the volume of solution (ml),
A = the measured absorbance,
$A^{1\%}_{1cm}$ = the specific absorption coefficient, i.e. the absorbance of a solution of 1 g of that carotenoid in 100 ml of solution.

Its values for some common carotenoids have been given[4]. Its more extensive tables have been published.[5, 11, 7, 8]

However, an arbitrary value of 2500 is often taken when no experimentally determined value has been reported, for an unknown compound, or to give an estimate of the total carotenoid content of an extract. It is difficult to achieve complete solution, especially of crystalline carotenoids even in favourable solvents, so carotenoid contents are easily underestimated. Also, there is doubt about the accuracy of some of the published $A^{1\%}_{1cm}$ values.

(ii) ***Quantitative determination by HPLC*** This is the most sensitive, accurate and reproducible method for quantitative analysis of carotenoids, especially when the instrumentation includes automatic integration facilities for measuring peak area. The relative amounts in each component in the chromatogram can be determined, provided the peak area can be calculated for each component at its λ_{max} by use of a multi-wavelength detector. For the determination of absolute amounts of concentration, calibration is necessary and this can be achieved by means of a calibration graph or by use of an internal standard.

GROUP B VITAMINS

Chemical Estimation of Thiamine

Bound thiamine of foods is released by incubation with the enzyme takadiastase and papain at pH 4.2. The crude extract is purified by passing through a column of activated decalso (or its equivalent). The thiamine in eluate is estimated by oxidising it to thiochrome and measuring the fluorescence. Lead acetate is used if decalro is not available. The results arc practically comparable.

Reagents

(i) ***Activated decalro or equivalent:*** Either of these should pass through 180 to 250-micron sieve successively once with hot 3% acetic acid, once with hot KC1 solution and again with hot 3% acetic acid followed by several washings with hot water. Each washing means stirring decalso in the liquid for 15 minutes, letting it settle and then decantation. The final wash solution should be free of chlorides when tested with 1% $AgNO_3$ solution. Thus treated decalso is stored under water in a stoppered bottle or in a dry state, dried at 100°C.

(ii) ***Na Acetate buffer (020M,pH4.2):*** 34 g of CH_3 COONa.$3H_2O$ is dissolved in 250 ml H_2O. 30 ml of this solution + 70 ml of 1 N acetic acid is diluted to 500 ml with H_2O.

(iii) ***Enzyme Solution:*** 150 mg of takadiastase + 75 mg of papain are suspended in 5 ml of (ii) above.

(iv) ***KCl Solution:*** 25% in 0.1N HCl (w/v).

(v) ***Isobutyl Alcohol:*** Redistilled (in an all glass apparatus). 105 to 108°C distilled fraction is collected and saturated with water.

(vi) ***Bromocresol green indicator:*** 0.4% in 70% ethanol.

(vii) ***K3Fe(CN)$_6$ solution:*** 1% in water.

(viii) ***Oxidising reagent:*** 1 part of (vii) + 9 parts of 15% KOH solution (w/v) in water.

(ix) ***Blank reagent:*** 9 parts of 15% KOH solution (w/v) in H_2O + 1 part water.

(x) ***Stock Thiamine solution:*** 25 mg of pure dry crystalline thiamine hydrochloride is dissolved in 250 ml of 0.1 N HCl..

1 ml = 100 ug.

Extraction The edible portion of the food is cut into small pieces, homogenized well in a blender with a suitable volume of water. After weighing the total slurry, a weighed aliquot = 15 ± 5 g of food is used for thiamine extraction. For dry foods 15 ± 5 g is weighed and used as such.

2 aliquots of the slurry or powder are taken into 2,250 ml flasks marked "Test" and "Recovery", 5 ml of (iii) + 80 ml (ii) added to each flask. To the "Recovery" flask is added diluted thiamine standard solution = 25 µg of thiamine HCl. About 2.5 ml of toluene is added to both the flasks, both incubated at 37°C overnight. At

the end of incubation period the enzyme is inactivated by heating them in a boiling-water bath, the volume made up to 100 ml with water, contents centrifuged on filtered.

Purification Two chromatographic columns, each column (25 × 1 cm), are marked separately as before "Test" (T) and "Recovery" (R) column. Small pieces of glass wool are placed at the bottom of T and R, then filled with activated decalso to a height of 8 cm and washed with 3 ml of 0.5% acetic acid.

20 ml of test and recovery are taken in beakers, their pH adjusted to 3.5 with IN HCl. The end point is checked in pH meter or with bromocresol green indicator, (yellowish green colour).

The samples and the washings are poured on the respective columns, the liquid is let to drain. The T and R columns are washed with 2 × 10 ml portions of boiling water, the eluates are also discarded.

Thiamine is eluated from the column successively with 10, 10 and 5 ml of boiling KCl solution. The eluates are collected directly in a 25 ml volumetric flask or in a conical flask, the volume made up to 25 ml after cooling.

In the absence of decalso the following method is used

Basic Pb acetate solution 180 g of lead acetate is dissolved in about 700 ml water through heat application. 110 g of finely powdered lead-oxide (litharge) is added to the hot solution, the boiling continued for 45 minutes with constant stirring. The mixture is cooled, filtered, the filtrate diluted to 1 litre.

10 ml of above solution is added to 20 ml of the food extract, mixed and centrifuged. To 20 ml of supernatant, 3 ml of 30% (v/v) H_2SO_4 and 17 ml water added. The precipitate is removed by centrifugation. The supernatant solution is stored at 5°C.

Here 30 μg of thiamine HCl is employed for recovery at the time of incubation with the enzyme.

Procedure

This involves conversion to thiochrome. 2 × 5 ml aliquots, 10 ml in case of extracts obtained with lead acetate method are taken from the purified extract T into 2 separating funnels each of 50 ml capacity or glass stoppered test tubes labelled T and B. Similarly, in a third tube marked R is taken 5 ml of the purified extract R-10 ml in lead acetate method. 1 ml of blank reagent (ix) is taken in the tube B, followed within 1 minute by 15 ml of (v). To the T&R tubes/funnels is added 1ml of (viii) + 15 ml of (v). Both the tubes/funnels are shaken vigorously for 1 minute, layers let to separate. The upper butanol layer is transferred to another test tube/funnel. This contains thiochrome. A small quantity of anhydrous Na_2SO_4 is added to remove traces of water. The fluorescence of the clear extract is read using suitable filters for thiamine estimation. A suitable reagent blank and a standard containing 2-5 ug of thiamine HCl should also be prepared to balance the fluorometer.

Calculation

μg thiamine/g of food

$$= \frac{\text{Test reading} - \text{blank reading}}{\text{Recovery reading} - \text{Test reading}} \times \frac{\text{Dilution factor}}{\text{Wt. of test material}}$$

The dilution factor in the decalso method would be 25 and in the lead acetate method 30.

B. Microbiological estimation The microorganism *Lactobacillus fermenti-36,* National Collection of Type Culture (NCTC No.6991) corresponding to American Type Culture Collection, ATCC No. 9338 has a specific requirement for its growth. The growth response on a defined medium complete in all respects except for thiamine is proportional to the concentration of the vitamin added in the medium, up to a certain stage using *L. fermenti-36,* direct titration of lactic acid formed should not be employed because the organism responds to pyrimidine and thiazole moieties of the molecule after 18 hours of incubation.

Reagents

1. The principal reagent is a basal medium prepared by mixing several stock solutions in definite proportions. The stock solution should be preserved under refrigeration in the dark with 0.1% $CHCl_3$ + 0.5% toluene added to prevent microbial growth.

2. Acid hydrolysed Vitamin-free Casein

(a) ***Extraction of Casein*** 100 g of vitamin-free casein is stirred with 250 ml of 95% ethanol for 15 minutes in a one litre beaker and filtered with suction aid. Extraction is continued using 250 ml lots of the ethanol to remove any residual vitamin from the casein.

Note: Commercial brands of vitamin-free casein contain some B complex vitamins which may give high blanks in the absence of above treatment.

(b) ***Hydrolysis*** The alcohol-washed casein is transferred into a round-bottomed flasks of 1 litre capacity, preferably with 2 necks, ground to a standard taper. It is mixed with 250 ml HCl (37%) and 250 ml water. The contents are refluxed for 10 ± 2 hours through a water cooled condenser over low heat to avoid frothing-an inherent property of casein.

Maximum possible content of HCl is removed by distillation over vacuum/reduced pressure. The resulting hydrolysate is concentrated to a thick paste. Air may be inducted through a bleeder tube to decelerate bumping during the final stages of concentration. 75 ± 5°C is the recommended temperature. The left over acid concentration should be sufficiently low lest subsequent neutralization may yield high salt concentration capable of retarding bacterial growth on the basal medium. These conditions are achieved by redissolving the paste in about 200 ml water, distillation repeated, contents reconcentrated. This removes excess HCl.

(c) ***pH Adjustment*** The hydrolysate paste is dissolved in about 700 ml water, pH adjusted to 3.5 with 40% NaOH. The contents colour is adsorbed on activated charcoal, 200 g at room temperature by stirring till the supernatant becomes light straw. The time needed depends upon the quality of charcoal used. Besides decolourisation, residual niacin or folic acid as a carry over from casein is also removed. Contents are filtered quickly either through a large fluted filter paper or by suction. The filtrate's pH is adjusted to 6.8, contents diluted to 1 litre, filtered if needed and finally stored under toluene and over $CHCl_3$ in the refrigerator. Occasionally during standing tyrosine precipitates, if so it is shaken up, the suspended material as well as fluid should be used. The insoluble material gets dissolved while preparing the final medium.

3. ***Adenine-Guanine-Uracil (AGU) solution*** 100 mg each of adenine sulphate, guanine HCl and uracil is heated in a 250 ml Erlenmeyer flask containing about 75 ml H_2O and 2 ml HCl. After complete dissolution, contents are cooled, otherwise the precipitate is dissolved by adding a few drops of conc. HCl and heating. This step is repeated till no precipitate is formed on cooling. The final solution is transferred to 100 ml volumetric flask and the volume made up.

4. ***Thiamine-Free Yeast supplement*** 20 g of Difco or equivalent yeast extract is dissolved in 200 ml of 0.5 N NaOH solution, contents autoclaved at 120-123°C for 30 minutes. The resulting solution is neutralized with glacial acetic acid and again autoclaved for 10 minutes to coagulate and precipitate any protein. Contents are filtered, pH of the filtrate adjusted to 1.5 with HCl using thymol blue as external indicator. The solution is shaken with 20 gms of activated charcoal for 20 minutes, filtered, pH of the filtrate adjusted to 1.5 if essential. The filtrate also is shaken like above, filtered, filtrate neutralized with NaOH, contents preserved under S-free toluene in a refrigerator. Its shelf life is about 3 months.

5. ***Photolyzed Peptone*** 40 g of Difco/Bacto or equivalent peptone is dissolved in 250 ml H_2O + 20 g NaOH in 250 ml water. Contents are kept aside for 24 hours in the room, then exposed to a light of 100-watt bulb kept at a distance of about 45 cms for 18 hours. After exposure the solution is neutralized with glacial acetic acid (25-28 ml), 14 g of Na_2SO_4 added, the solution made up to 800 ml with water and preserved under S-free toluene in the refrigerator.

6. ***Salt solution A:***– 25 g of K_2HPO_4 + 25 g of KH_2PO_4 dissolved in H_2O then diluted to 500 ml with H_2O. It also is stored under S-free toluene in a refridgrater.
7. ***Salt solution B:***– 10.0 g of $MgSO_4.7H_20$, 0.5g each of NaCl, $FeSO_4.7H_2O$ and $MnSO_4.4H_2O$ in H_2O and diluted to 500 ml. 5 drops of conc. HCl are added and then stored as at (6).
8. ***Stock Thiamine solution (1000 μg/ml)***– 100 mg of thia mine is dissolved in a little water, the volume made up to 100 ml

with 2% (v/v) HCl. This solution is stored in the refrigerator. It keeps well for several weeks.

9. ***Thiamine working standard (0.02 μg/ml)*** For various assays, first 1 ml of (8) is diluted to 100 ml with H_2O to get a solution of 10 ug/ml. 5 ml of this solution is diluted to 1000 ml to obtain the working standard.

10. ***Basal medium*** The thiamine-free basal medium for 100 tubes should contain,

(5) above	200 ml
(2), (11) above	25 ml each
Glucose	20 g
Hydrated Na acetate, (3),(4)	10 g each
NaCl	5 g
Salt solution A, B	5 ml each

All these are mixed, pH adjusted to 6.5 with NaOH solution and the volume made up to 500 ml with water. If needed this solution may be filtered through a suitable sintered glass funnel or filter paper.

11. ***Enriched liver tryptone agar medium for stock culture*** The following ingredients are mixed in glass distilled water, volume made up to 100 ml.

Glucose	1.0 g
Difco/Bacto or equivalent tryptone	1.0 g
$K_2H PO_4$	0.2 g
$CaC0_3$	0.3 g
Liver extract	10 ml
Salt solution A,B	0.5 ml each
Thiamine	100 mg
Agar Difco/Bacto or equivalent	1.5 to 2.0 g is dissolved by steaming.

Aliquots are distributed into bacteriological tubes, tubes are plugged with cotton and sterilized at 115°C for 10 minutes. Test for sterility is conducted by incubation of the tubes at 37°C for 24 to 48 hours, then stored in a refrigerator.

12. ***Liver extract for reagent*** 10.5 kgs of fresh liver is ground, suspended in 2 litres of water and heated on a steam-bath for one hour and filtered through cheese cloth. The filtrate is neutralized

to pH 7.0, again heated for 15 minutes, filtered through coarse filter paper, the filtrate is stored in a dark bottle under toluene in the refrigerator.

13. Culture medium for growing inoculum 5 ml each of basal medium (10 above) into clean and dry bacteriological tubes. 10 ug of thiamine is added to each tube, volume made up to 10 ml with glass-distilled water, plugged with cotton and sterilized in an autoclave at 120-123°C for 10 minutes, cooled and stored in the refrigerator.

14. Isotonic salt solution 900 mg NaCl is dissolved in 100 ml water. 10 ml is distributed in each tube, each tube is plugged with cotton and sterilized at 120-123°C for 15 minutes.

PROCEDURE

(a) Preparation of stock culture Stab culture is prepared in 2 or more stock agar culture tubes using a pure culture of *L. fermenti-36.* It is incubated for 16-17 hours at 37.0°C ± 0.5°C. It is stored in the refrigerator for not longer than one week before transferring to the new stab.

(b) Preparation of inoculum Cells from (a) are aseptically transferred into (13) with a sterile Pt. needle, incubated like (a) above, centrifuged, the supernatant is poured off aseptically. The cells are again suspended in 10 ml of (14), centrifuged again after mixing and pouring off the supernatant. The cells are resuspended into another 10 ml of sterile saline. Its few drops are diluted with 10 ml sterile saline. This diluted inoculum is used immediately for inoculating assay tubes.

(c) Extraction for assay The test material is finely powdered/ground/homogenised and a proportionate amount is weighed, which should contain 5-10 ug of thiamine, into 120 or 125 ml Erlenmeyer flask containing 25 ml of 0.1N H_2SO_4. Acid volume is increased if needed when the material volume is large. The flask is plugged with cotton and steamed for 30 minutes. After cooling, its pH is adjusted to 4.5 with 2.5M Na-acetate buffer, using bromo-cresol green as external indicator. 20 mg each of takadiastase and pepsin are added and mixed well, then incubated under a thin layer of S-free toluene for 21 ± 3 hours at 37°C. The digest is kept in a steamer for 30 minutes at 100°C to stop enzymatic action and to drive off toluene. Contents are cooled, volume made up to 100 ml with water, mixed and filtered through Whatman No.l filter paper or equivalent. The filtrate is kept in the refrigerator.

(d) Preparation for assay pH of an aliquot of the sample extract is adjusted to 6.5 with NaOH, the volume made up to 50 ml with water so that 1 ml = 0.015 µg of thiamine (approx). This is sample extract I.

Another aliquot is taken in a 100/125 ml Eslenmeyer flask and 4.0 ml of freshly prepared 10% solution of crystalline sulphite is added to it. Its pH is adjusted to 5.4 with 1 N H_2SO_4, using a suitable external indicator. The volume is made up to about 30 ml, it is autoclaved at 120°-123°C for 15 minutes, cooled, excess Na_2SO_3 is oxidised with a measured amount of 33% H_2O_2 using a freshly prepared mixture of equal parts of 5% Kl solution, 1% starch solution and 50% H_2SO_4 as external indicator. The volume is made up to 50 ml. This sulphite treated extract is sample extract II and serves to correct for any turbidity due to sample extract.

(e) Preparation of assay tubes 15 uniform size tubes are serially numbered. 5 ml of basal medium (10) is added into each of the 15 tubes followed by 2 ml each of sample extract I to first five tubes. 2 ml each of sample extract II is added to next ten tube. 3 ml of water is added to each of the first 6 tubes, i.e. 5 tubes with sample extract I and one with, sample extract II. The 6th tube serves as the blank for turbidimetric reading.

Standard thiamine solution (9) is added into the remaining 9 tubes at levels of 0.02 ug, 0.03 ug and 0.04 ug/tube, thus each level having been added in triplicate. The volume of these 3x3=9 tubes is made up to 10 ml. These serve as standards. All tubes are either plugged with cotton or capped with aluminium caps and sterilized at 115°C for 10 minutes, cooled and readied for incubation.

(f) Assay procedure The assay tubes (e) above are brought to 37°C, inoculated aseptically with a drop each of the diluted inoculum/ tube, incubated for 17 to 18 hours at 37° ± 0.5°C. The tubes are cooled at the end of incubation period for 15 minutes at 4° to 5°C, the growth is measured nephelometrically, using tube No.6 as the blank.

Calculation

The average of three readings for each concentration is plotted graphically against ug of thiamine used. The average value of the vitamin content of the replicates is determined by interpolation from the readings,

$$\mu\text{g diamine/g of food sample} =$$

$$\frac{\mu\text{g of thiamine/2ml of Extract I} \times 25}{\text{Aliquot of sample taken}} \times \frac{100}{\text{weight of sample taken}}$$

ESTIMATION OF RIBOFLAVIN IN FOODS

To repeat again, it is ensured that there is no deterioration of riboflavin during assay. Powders and liquids are mixed thoroughly to achieve homogenity. Dry materials like bread, biscuits and grains are ground to a fine mesh. Butter is melted under constant stirring. Margarine/cheese, etc. test materials should contain portions of the surface as well as of the interior. Wet or fresh materials are minced with a knife or scissors or homogenised in a blender, if needed, in the presence of the extracting solvent.

Riboflavin can be estimated by chemical as well as a microbiological mediod.

1. ***Chemical method*** Riboflavin fluoresces when exposed to light of wave length 470 ± 30 mμ. The intensity of the fluorescence is proportional to the concentration of riboflavin in dilute solutions.

Riboflavin phosphate and flavin adenine dinucleotide exhibit the same characteristic yellow colour and yellow fluorescence as riboflavin. The riboflavin is measured in terms of difference between the fluorescence before and after chemical reduction by hydrosulphite which will reduce riboflavin and its co-enzymes to colourless compounds which do not fluoresce.

The fluorometer should be suitable for accurately measuring fluorescence of solutions containing riboflavin in concentrations of 0.05 to 0.20 μg/ml. A fluorometer having the input filter of narrow transmittance with maximum of about 440 mu and the output filter of narrow transmittance with a maximum of about 565 m)i has been found satisfactory.

Reagents

1. ***Standard HCl (0.1 N)***
2. ***NaOH solution: 4% (w/v)***
3. ***Dil HCl:1:1 (v/v)***
4. ***Riboflavin stock solution I*** 50 mg of USP riboflavin reference standard/IP standard previously dried and stored in dark in a desiccator over P_2O_5,. is added to about 300 ml of 0.02N acetic acid, the mixture is warmed on a steam bath with constant stirring until the riboflavin is completely dissolved. Contents are cooled, 0.02N acetic acid is added to make the volume to 500 ml. This solution is stored under toluene in the cold in a dark bottle.

   1 ml = 100 ug of riboflavin

5. ***Riboflavin stock solution II*** 50 ml of (4) above is diluted to 500 ml with 0.02 N acetic acid and stored like (4) above.

   1 ml = 10 ug of riboflavin

6. ***Standard riboflavin solution*** 10 ml of (5) is diluted with water to make 100 ml.

1 ml = 1 ug of riboflavin

This solution must be prepared afresh for each assay.

7. ***$KMn0_4$ solution*** 4 g of $KMnO_4$ crystals in 100 ml water, kept for a few days for osmatic homogenisation, filtered and stored in a dark bottle.
8. ***H_2O_2 3%***
9. ***Na hydrosulphite (CP)*** It should be unexposed to light or air.

Caution Throughout the procedure, pH of the solution must be less than 7.0 to prevent riboflavin loss.

Sample preparation A weighed quantity of the sample is taken in a flask of suitable size. HCl (1) = ml not less than 10 times the dry weight of the sample in gms is added to have the resulting solution of less than 0.1 mg/ml of riboflavin. Readily insoluble sample materials are comminuted to facilitate even dispersion in liquid. The solution is vigorously agitated, the sides of flask are washed down with HCl (1). The mixture is heated in an autoclave at 122 ± 1°C for 30 minutes then cooled to room temperature. Dumping is removed by mixture agitation till even dispensability is achieved. Agitation is vigorously done otherwise also to adjust mixture pH between 6.0 to 6.5 with NaOH. Dil HCl (3) is immediately added until no further precipitation occurs. The mixture is diluted to a known volume so that 1 ml of the final mixture contains more than 0.1 ug of riboflavin. The mixture is filtered through Whatman No. 1 or equivalent filter paper immune to riboflavin absorption. If filtration is not easy, it is centrifuged and then filtered through fritted glass, using suitable filter aid which may often be substituted for or proceeded by filtering through a filter paper. An aliquot of the clear filtrate is checked for dissolved protein by adding dropwise, first dil. HCl(3) and if no precipitate occurs then NaOH solution (2) with vigorous agitation and proceeded as described below.

(a) If no precipitation occurs then pH of the solution is adjusted to 6.8 by adding NaOH solution (2) accompanied by vigorous agitation. The solution is diluted to a known volume so that 1 ml of the solution contains about 0.10 ug of riboflavin. In case of cloudiness it is refiltered. This is solution A.

(b) If further precipitation occurs, the solution is readjusted to have maximum precipitation. Like before the solution is diluted to a known volume so that 1 ml has riboflavin more than 0.10 ug. An aliquot of the clear filtrate is proceeded as described above. Let this be solution B.

Estimation

(a) To each of the two tubes/reaction vessels is added 10 ml of the sample solutions. If fluorometer needs tubular cuvettes, then all reactions are carried out in matched sets of these cuvettes. To one of these tubes is added 1 ml of (6) and mixed, to the other 1 ml of water is added and contents mixed. 1 ml of acetic acid is mixed followed by addition with mixing 0.5 ml of (7) whose quantity may be increased for sample solutions that contain excess of oxidizable material, but not more than 0.5 ml in excess of the required, for complete oxidation of foreign matter. Two minutes are enough for this reaction. 0.5 ml of (8) is added and mixed in both the tubes, where upon the permanganate colour gets destroyed within

10 seconds. The tubes are vigorously shaken until excess oxygen is expelled. Gas bubbles remaining on the tubes sides after cessation of foaming are removed by tipping the tubes so that solution flows slowly from end to end. Frothing can be tackled by adding a drop of alcohol/acetone or n-octanol.

(b) Fluorescence (A) of the sample solution containing the added 1 ml of (6) is measured, so also of the other tube containing water (B). To this tube (B) is added and mixed 20 mg of powdered (9). Fluorescence of this, say (C) is noted within five seconds.

Note: Sodium hydrosulphiate should be of high purity and kept from undue exposure to light or air. A quantity appreciably in excess of 20 mg may reduce foreign pigments and fluorescing substances or both thereby causing erroneous results. Suitability of Sodium hydrosulphite may be checked as follows.

To each of 2 or more tubes, 10 ml of water and 1ml of riboflavin solution containing 20 μg of riboflavin/ml is added and proceeded as (a) above with respect to the addition of acetic acid, $KMnO_4$ and H_2O_2 solutions, then while adding with mixing 8 mg of Na hydrosulphite, the riboflavin should be completely reduced in less than 5 seconds.

Calculation

$$\text{mgm of riboflavin/ml of the final solution} = \frac{B-C}{A-B} \times \frac{1}{10} \times \frac{1}{100}$$

(value of B-C/A-B shall be not less than 0.66 and more dian 1.5); riboflavin content mg/100 g on dry/wet basis, as the case may be.

II. *Microbiological method* The microorganism *Lactobacillus casei* NCTC NO 6375 or ATCC NO 7469, has a specific requirement for riboflavin for its growth and lactic acid production. The growth response on a defined medium complete in all respects, except the vitamin under test, is proportional to the concentration of the vitamin added to the medium up to a certain range.

Note: NCTC connotes National Collection of Type Cultures and ATCC connotes American Type Calcutta Collection.

Apparatus

(a) ***Incubator*** Maintaining uniform temperature in the range 30°-37°C (± 0.5°C). A water-bath also serves well.

(b) ***Bacteriological tubes (rimless)*** **15 × 150** mm or **25 x 200** mm.

(c) ***Culture tube racks*** to hold vertically a total of 120 tubes and so designed to permit free air circulation and of non-rusting materials.

Reagents

The principle reagent is a basal medium prepared by mixing a number of stock solutions and chemicals in definite proportion. The stock solutions must be preserved under refrigeration in the dark with 0.1% $CHCl_3$ and 0.5% toluene to prevent microbial growth. The solutions must be protected from light, particularly the vitamin solutions, and extracts of food materials to be assayed.

(i) ***Salt solution A*** 25 g each of K_2HPO_4 and KH_2PO_4 is dissolved in water and the volume made up to 500 ml with water. It is stored under toluene.

(ii) ***Salt solution B*** 500 mg each of NaCl, $FeSO_4.7H_2O$ and $MnSO_4.4H_2O$ + 10 g of $MgSO_4.7H_2O$ are dissolved as well as diluted with water to 500 ml. 5 drops of conc. HCl are added and the solution stored under toluene.

(iii) ***Dil HCl, 0.1N approx*** 8.5 ml of cone HCl is diluted to 1 litre with water.

(iv) ***Stock riboflavin solution A*** 25 μg riboflavin/ml in 0.02 N acetic acid. Its procedure is similar to the one adopted in chemical method.

(v) ***Stock riboflavin solution B*** 10 μg/ml in 0.02 N acetic acid, prepared as in the chemical mediod.

(vi) ***Riboflavin working standard*** 0.1 μg/ml in 0.02 N acetic acid, prepared as in the chemical method.

(vii) ***Photolyzed peptone*** 40 g of Difco/Bacto or equivalent peptone is dissolved in 250 ml of water and 20 g of NaOH in 250 ml of water. The two solutions are mixed, kept for 24 hours in the room but exposed to a 100 watt bulb kept at a distance of 45 cms for 18 hours to ensure destruction of all riboflavin in the peptone. At the end of the exposure, the solution is neutralized with glacial acetic acid 25 to 28 ml, 14 g of anhydrous Na-acetate is added. The volume is made up to 800 ml with water and preserved under a layer of sulphur free toluene. Shelf life is usually 14 days under refrigeration, however if precipitation occurs or the solution becomes cloudy before this time, it must be discarded.

(viii) ***Cystine solution*** 0.7% (w/v) 1 g of 1-cystine is suspended in 20 ml of water. Conc. HCl is added dropwise about 10 ml until the crystals dissolve. The volume is made up to 1 litre with water, the final solution is preserved under sulphur-free toluene.

(ix) ***Yeast supplement (riboflavin free)*** 100 g of Difco/Bacto or equivalent yeast extract or autolyzed yeast is dissolved in 500 ml of water. 150 g of basic lead acetate is dissolved in 500 ml of water. The two solutions are mixed, the pH of the resultant solution is adjusted to approx. 10 with conc. NH_4OH using an external indicator. The solution is filtered through a Büchner funnel, filtrate's pH adjusted to about 6.5 with glacial acetic acid and H_2S passed to remove lead as PbS. The black precipitate is filtered, the volume of the filtrate is adjusted to 1000 ml. Excess H_2 S is removed by passing air for sometime or removing 200 ml of water under vacuum. This is preserved with toluene and $CHCl_3$, under refrigeration.

(x) ***Basal medium stock solution for 100 tubes*** The following are mixed,

(a)	Photolyzed peptone	100 ml
(b)	0.1% cystine	100 ml
(c)	yeast supplement solution	20 ml
(d)	salt solution A	10 ml
(e)	salt solution B	10 ml
(f)	Glucose, anhydrous	10 g

It is dissolved in the mixture of the solution, pH adjusted to 6.8 with IN NaOH, the volume made up to 500 ml with water.

(xi) ***Enriched agar medium for stock culture*** 2 g of anhydrous glucose, 1 g of peptone (Difco/Bacto or equivalent), 100 m[illegible] of cystine, 1 ml each of salt solution A and B, all to[illegible]

are dissolved in 150 ml of water, pH adjusted to 6.8, the volume made up to 200 ml with water. 3.5 g of Bacto/Difco or equivalent agar is added to the mixture, it is steamed till agar gets dissolved. This is dispensed into 20 test tubes, each tube stoppered with cotton plugs and sterilized at 120°-123°C for 15 minutes.

(xii) ***Culture medium for growing inoculum*** 5 g of peptone, 1 g of yeast extract, 10 g each of anhydrous glucose and Na acetate, 5 ml each of salt solutions A and B, together dissolved in 200 ml water in a beaker. pH is adjusted to 6.8 with 0.1N NaOH, contents diluted to 500 ml, filtered, 10 ml is added to test tubes, plugged with cotton and sterilized at 120-123°C for 15 minutes.

(xiii) ***Isotonic salt solution*** 900 mg NaCl in 100 ml of water. 10 ml is added to each tube, cotton plugged and sterilized as before.

(xiv) ***NaOH (1N) standardised*** 40 g NaOH pellets dissolved in 1 litre of H_2O and standardised against potassium hydrogen pthalate of known strength.

(xv) ***NaOH 0.1N standardised***

(xvi) ***Bromothymol blue indicator solution*** 1 g indicator is weighed into a small beaker. 1.6 ml of 0.1N NaOH is added, the powder is stirred with a glass rod till dissolved, then diluted to 250 ml with water. This solution may also be prepared by dissolving in a few ml 95% ethanol adding 1.6 ml of above alkali, then diluting to 250 ml with water.

Procedure

Active stab culture of the stock *L. casei* not older than 14 days for the preparation of inoculum.

(a) ***Preparation of stock culture*** Stab cultures are prepared in 2 or more agar stock culture tubes (xi). A pure culture obtainable from National Collection of Type Cultures/American Type Culture Collection should be used. These are incubated for 20 $\pm$ 4 hours at 37.0 $\pm$ 0.5°C and stored in the refrigerator under aseptic conditions for not longer than a week before transferring to a new stab.

(b) ***Preparation of inoculum*** The culture cells from stock culture are transferred aseptically to a sterile tube of inoculum culture medium (xii). This culture is incubated for 15 ± 3 hours at 37°C. Inoculum older than 24 hours is not fit for assay. In case of emergency, the inoculum may be refrigerated for 24 hours. The cotton plug is secured with rubber band or adhesive tape, the contents are centrifuged for about 15 minutes. The supernatant liquid is decanted, the cells are suspended aseptically in 200 ml of sterile iso-tonic salt solution. Centrifuging may be repeated, so also suspension of cells in sterile salt solution if needed. A sterile syringe filled with resuspended cells is used at once. 20 ml of inoculum shall suffice for 200 assay tubes. Alternatively, a sterile pipette or a sterile glass tube drawn out to a capillary may also be used in place of a syringe.

(c) ***Preparation of sample-extraction and hydrolysis*** A homogeneous sample containing 10 mg or more of riboflavin is weighed in a 100/125 ml conical flask, 50 ml of (iii) is added, contents auto-claved for 15 minutes at 120-123°C, cooled to room temperature, pH adjusted to 4.5 and transferred to 100 ml volumetric flask. The volume is made and filtered. 50 ml/aliquot containing about 5 ug of riboflavin of the filtrate is measured into 100 ml volumetric flask, pH adjusted to 6.8, contents diluted to 100 ml. 5 levels of sample extract ranging from 0.05 to 0.25 ug are used; each tube varying by not less than 0.5 ml between two serial conditions.

(d) ***Preparation of standard tubes*** In duplicate tubes serially numbered are added aliquots of (vi), volume made up in each tube to 5.0 ml by water addition and 5 ml (x) contents mixed well by rotating in the palm of the hand. 5 ml of water and 5 ml of (x) is added into two tubes to serve as blanks. For establishing standard curve, 0.05,0.10,0.15,0.20 and 0.25 ug of riboflavin and automatic pipettes are recommended. Each tube is covered with cotton plug/ aluminium cap and sterilized at 115°C for 10 minutes. The tubes should be cooled in the dark before inoculation.

(e) ***Preparation of assay tubes*** To duplicate tubes, numbered or marked serially is first added at five levels of aliquot 0.5 to

5.0 ml of the test extract of food material (approximately within 0.05 to 0.20 (xg) then sufficient water to bring the volume to 5 ml and 5.0 ml of (x). Each tube is mixed by rotating in the palm of the hand, plugged with cotton or covered with aluminium caps and sterilized at 115°C for 10 minutes.

(f) ***Inoculation and incubation*** All tubes are cooled to the incubation temperature because small differences in temperature of the various tubes at the start of the assay influence the rate of growth. Each tube is aseptically inoculated with one drop of the inoculum (xii) in a chamber or hood. The tubes are incubated at 37°C for about 72 hours at a uniform temperature.

(g) ***Titration*** Contents of each tube are transferred into 100/125 ml conical flasks marked/numbered correspondingly. Each tube is rinsed twice with about 10 ml of water, rinsing also added respectively. About 0.2 ml of (xvi) is added, contents titrated with (xv) to a green colour, pH about 6.8. One flask is held for reference colour for 10 titrations and then substituted with another new flask to get the correct point. Blank is titrated to the same end point.

(h) ***Calculation*** A standard curve is drawn for the assay by plotting the ml of (xv) against concentration of standard vitamin added per tube, after subtraction of the value for blank.

The vitamin content of the tubes in the unknown series is determined by interpolation of their titre values on the standard curve. The vitamin content of test material is calculated from the average of the values for 1 ml of test solution obtained from not less than 3 sets of the tubes which do not vary by more than 10% from the average, using the formula:

ugm of riboflavin in food sample*

$$\frac{\text{Average } \mu\text{g/ml} \times \text{Volume} \times \text{dilution factor}}{\text{weight of sample in grams}}$$

* On dry basis/wet basis as the case may be.

NIACIN (NICOTINIC ACID)

Preparation of sample Powders and liquids are mixed thoroughly to have a well homogenised test material. Dry foods like bread, biscuits and grains are ground to a fine powder of less than 500 μm size. Butter is melted under constant stirring. Margarine/ cheese/ other such foods samples are made from portions of the surface as well as of the interior. Wet or fresh material is mixed with a knife or scissors, alternatively homogenised in a blender, if necessary, in the presence of the extracting solvents.

Chemical method Nicotinic acid reacts with cyanogen bromide to give a pyridinium compound which undergoes rearrangement yielding derivatives that couple with aromatic amines, giving coloured compound which are estimated quantitatively photometrically.

Reagents

(i) ***Nicotinic add (NA) stock solution*** 50 mg of USP nicotinic acid reference standard or equivalent I.P. standard, previously dried and stored in dark in a desiccator over P_2O_5, is dissolved in alcohol and the volume made up in a 500 ml volumetric flask with alcohol. This solution is stored at 10°C in a dark bottle.

1 ml = 100 μg of nicotinic acid (NA)

(ii) ***NA working standard solution*** 2 ml of (i) is transferred to a 50 ml volumetric flask, brought to ambient temperature then diluted to volume with water.

1 ml = 4 μg of NA.

(iii) ***(a) Sulphanilic acid solution 10% (w/v)*** 20 g of sulphanilic acid + 170 ml of water is dissolved by adding NH_4OH solution. The pH is adjusted to 4.5 with dil HCl (1:1) (v/v), using bromocresol green as indicator, volume made up to 200 ml colourless solution.

(b) ***Sulphanilic acid solution 55% (w/v)*** 55 g of sulphanilic acid is dissolved by shaking with 27 ml each of H_2O and NH_4OH solution. Contents may be warmed if warranted. pH of this solution is adjusted to 7 with a few drops of NH_4OH/dil HCl acid. The final volume is made to 100 ml and stored in dark.

(iv) ***Cyanogen bromide solution (CNBr) 10% (w/v)*** 40 g of freshly prepared CNBr is shaken with 370 ml of warm water

(40°C), to dissolve it. The contents are cooled and diluted to 400 ml taking care not to allow CNBr or its solution to come in contact with skin. The solution is prepared under a hood. Alternatively, CNBr can be prepared by mixing Br_2 water and KCN solution, in slight excess.

(v) ***Dil NH_4OH Solution*** 5 ml of 28% NH_4OH is diluted to 250 ml with water.

(vi) ***Dil HCl/HBr, 1:5 (v/v)***

(vii) ***Phosphate Buffer solution*** 60 g of $Na_2HPO_4.7H_20$ + 10 g of KH_2PO_4 are dissolved in warm water and diluted to 200 ml.

Procedure

(I) *(a) Extraction of Niacin from non-cereal foods and feeds and Preparation of sample solution* 40 g of sample is weighed into a 1 litre Erlenmeyer flask, 200 ml of $(NH_4)_2SO_4$ is mixed, contents autoclaved for 30 minutes at 120°C to 123°C, cooled, their pH adjusted to 4.5 with standard NaOH solution using bromocresol green solution as external indicator, diluted to 250 ml with water and filtered.

17 g of $(NH_4)_2SO_4$ is weighed in a 50 ml volumetric flask, 40 ml aliquot sample solution is pipetted in it, volume made up with water; contents shaken vigorously, filtered, filtrate homogenised. 1 ml aliquot and 0.5 ml aliquot with 0.5 ml water is used for colour development.

(b) ***Preparation of standard solution*** 40 ml of (ii) + 17 g $(NH_4)_2SO_4$ in a 50 ml volumetric flask, the volume made up with H_2O.

1 ml = 3.2 μg of Niacin.

(c) ***Preparation of standard curve with standard vitamin*** See Table 9.1.

Table 9.1.

(1)	*(2)*	*Standard Blank (A)* *(3)*	*Standard Solution (B)* *(4)*
1.	Standard solution	lml(b)	1 ml (b)
2.	H_2O	5.0ml	nil
3.	(v) above	0.5ml	0.5ml

(1)	*(2)*	*Standard Blank (A)* *(3)*	*Standard Solution (B)* *(4)*
4.	(iv) above	nil	5.0ml
5.	(iii) (a) above	2.0ml	2.0ml
6.	(vi) above HCl	0.5ml	0.5ml
	Total	9.0ml	9.0ml

All solutions mentioned in Table 9.1 are added in a single tube, its colour read before proceeding with the next tube.

(i) Standard solution (b) above is taken in standard blank tube, while swirling the tube to impart a rotary motion in liquid, little (v) is added, swirling continued, (iii) (a) is added followed immediately by the addition of 0.5 ml of dil. HCl (vi) mixed again and placed in the photoelectric colourimeter. The instrument is adjusted at zero absorbance at any specific wavelength (usually 420 mu) within 30 seconds after the addition of (iii) (a),

(ii) The standard solution (b) above is treated in a standard solution tube in the same way as standard blank tube with respect to addition of (v). The tube is shaken immediately and solution (iv) added and contents reshaken. After 30 seconds the tube is again shaken and (v) is added, followed immediately by the addition of 0.5 ml dil. HCl (vi), mixed again and stoppered. Absorbance of standard solution in the photoelectrical colourimeter is read.

D. Determination ofNiacin in the Sample See Table 9.2.

Table 9.2.

	Sample Blank (C)	*Unknown Sample (D)*
1. Sample solution	1.0ml	1.0ml
2. H_2O	5.0ml	nil
3. (v)	0.5ml	0.5ml
4. (iv)	nil	5.0ml
5. (iii) (a)	2.0ml	2.0ml
6. Dil HCl (vi)	0.5ml	0.5ml
Total	9.0ml	9.0ml

Colour of sample blank (C) and unknown sample (D) is developed like (i) and (ii) above. With sample blank set at zero absorbance, absorbance of sample solution also is determined.

Calculation

μg of Niacin/g of food sample = 3.2 BC/AW,

where, B = absorbance of 1ml of sample solution,

C = total volume of sample extract,

A = absorbance for 3.2 mg of niacin and

W = weight in gm of the sample taken for vitamin extraction.

Preparation of samples and extraction of Niacin from Cereal Products

2.5 g of the sample containing 2.5 ug of niacin is taken in a 250 ml Erlenmeyer flask containing 1.5 gms of $Ca(OH)_2$ 90 ml of water is mixed thoroughly. The flask is autoclaved for 2 hours at 120° to 123°C. After autoclaving, the hot contents are mixed, cooled to 40°C, and the total volume made up to 100 ml and let to stand for decantation. 50 ml of this supernatant is taken and kept in an ice-bath for 15 minutes, then centrifuged and 20 ml of supernatant is collected in a centrifuge tube containing 8 g of $(NH_4)_2$ SO_4 and 2ml of phosphate buffer (vii) above. Contents are shaken till dissolved, warmed upto 55° to 60°C, centrifuged for 5 minutes and filtered through Whatman No. 12 filter paper or equivalent.

For the preparation of standard curve, only 0, 50,100,150,200 and 250 ug of Niacin is taken and proceeded as described in Table 9.3.

Table 9.3

S. No.	*Reagent*	*Standard Blank*	*Reagent Blank*	*Sample Solution*
1.	Standard Solutipn	5.0ml	nil	5.0ml
2.	CN Br (iv)	nil	10.0 ml	10.0 ml
3.	H_2O	10.0 ml	5.0ml	nil
4.	(iii) (b)	1.0ml	1.0ml	1.0ml
	Total	16.0 ml	16.0 ml	16.0 ml

After adding (1) and (3) reagents, all tubes are kept in ice-salt bath for 30 minutes. 10 ml of cold (iv) is added and within 30 seconds (iii) (b) is also added as indicated above, contents mixed thoroughly, the tubes again kept in ice-salt bath. The colourimeter is set at 100% transmittance with 470 mμ wave length with standard blank, optical density of the other tubes is read width 12 to 15 minutes after addition of (iii) (b).

Concentration of niacin versus optical density (od) (Y-X) is plotted where X is od for reagent blank and Y for standard solution.

Table 9.4
Determination of niacin in test cereal samples

S. No.	*Reagent*	*Standard Blank*	*Reagent Blank*	*Sample Solution*
1.	Sample solution	5.0ml	nil	5.0ml
2.	CNBr(iv)	nil	10.0 ml	10.0 ml
3.	H_2O	10.0 ml	5.0ml	nil
4.	(iii) (b)	1.0ml	1.0ml	1.0ml
	Total	16.0 ml	16.0 ml	16.0 ml

Repeat processes as done for the preparation of standard curve with sample and reagent blanks as well as sample solution from the standard curve, niacin concentration is found corresponding to (Z-X) od. Where Z is od of sample solution

Calculation

$$\mu\text{gm of niacin/g of cereal product} = \frac{100}{5 \times 2.5}(Z - X)$$

$$= 8(Z-X)$$

Microbiological Method

This method is based upon the observation that *Lactobacillus arabinosus* 17-5, NCTC No.6376, i.e. ATCC No.8014 requires niacin for growth. NCTC is the abbreviation of National Collection of Type Cultures and ATCC means American Type Culture Collection. Using niacin-free but otherwise complete basal medium, growth responses of the organism are compared quantitatively in standard and sample solutions. Either the acid or the turbidity produced by the organism is measured to determine the extent of growth and thereby the amount of niacin in the test solution.

Reagents

(i) ***Acid-hydrolysed casein*** 50 g vitamin free casein in a 500 ml, ground joint flask is refluxed with 250 ml of IN HC1 on a heating mantle for 12 hours. The solution is concentrated very slowly under reduced pressure when a thick paste of casein hydrolysate is left.

This paste is dissolved in about 200 ml H_2O and again concentrated under vacuum. The black residue is dissolved in 300 ml H_2O, pH adjusted to 3.5 using 40% NaOH solution. The hydrolysate is decolourised with activated charcoal, warmed and filtered. The decolourisation procedure is repeated 3-4 times till a straw coloured hydrolysate is obtained, pH adjusted to 6.8 and diluted to 500 ml and its solid weight determined

1 ml hydrolysate = 90 mg of solid

(ii) ***Cystine-Tryptophan solution*** 4.0 g of l (-) cystine and 1.0 g of l (-) tryptophan (or 2.0 g of dl-tryptophan) is suspended in 750 ± 50 ml water, heated to 75 ± 5°C, 20% HC1 added dropwise with stirring until the solids are dissolved. About 12 ml of 20% HC1 is enough. Contents are cooled to room temperature, the volume made up to 1 litre in a volumetric flask with water.

(iii) ***Adenine-Guanine-Uracil solution*** 100 mg each of adenine sulphate, guanine hydrochloride and uracil are heated in a 250 ml Erlenmeyer flask containing about 75 ml water + 2 ml of conc. HCl. When all solids get dissolved, the contents are cooled. In the event of precipitation, a few drops of conc. HC1 are added and contents reheated. This procedure is continued till no precipitate is formed on cooling. Only then the contents are transferred to a 100 ml volumetric flask and the volume made up with water.

(iv) ***d-Ca pantothenate-p-amino benzoic acid-pyridoxine HCl solution*** 10.0 mg each of d-Ca pantothenate p-amino-benzoic acid and pyridoxine HCl is transferred to a 1 litre volumetric flask and diluted to volume with water. Pyridoxine being photo labile, this solution is stored in the dark.

(v) ***Riboflavin-ThiamineHCl-Biotin solution*** 1 mg of xtalline biotin (free acid) in 100 ml of 0.02 N acetic acid, 4.0 ml (equivalent to 40 µg) of this solution is added to a 1 litre volumetric flask. 20 mg of riboflavin + 10 mg of thiamine HCl is added to the flask, and the volume made up with 0.02 N acetic acid. This solution is also stored in the dark to prevent destruction of riboflavin by light.

(vi) ***(a) Salt solution** A* 25 g of K_2HPO_4 + 25 g of $KH_2\ PO_4$ is dissolved in water men diluted to 500 ml with water.

(b) Salt solution B The following are dissolved in H_2O, then diluted to 500 ml like at (a) above.

$MgS0_4.7H_20$	10.0g
NaCl	0.5 g
$FeS0_4.7H_20$	0.5 g
$MnSO_4.4H_2O$	0.5 g

(c) 5 drops of conc. HCl is added and stored under toluene. NaCl may be omitted when HCl-hydrolysed casein is used.

(vii) ***Niacin stock solution*** (a) 50 mg of USP or equivalent IP niacin is weighed in 500 ml of 50% ethanol and stored under refrigeration.

1 ml = 100 µg

(b) Niacin working standard: 1ml of (a) above is dissolved in 1 litre of water.

1ml = 0.1 µg

(viii) ***Basal medium stock solution***	***For 100 assay tubes.***
Casein hydrolysate	50 ml
Cystine - tryptophan	50ml
Adenine-guanine uracil	10 ml
d-Ca pantothenate-p amino benzoic acid pyridoxine	10ml

Riboflavin - thiamine HCl-biotin	10 ml
Salt solution A	10 ml
Salt solution B	10 ml
Anhydrous glucose	10 g
Anhydrous Na acetate	10 g
H_2O	250 ml

All these ingredients are mixed in a 500 ml Erlenmeyer flask marked at 450 ml. pH of the contents is adjusted to 6-8 with 40% NaOH using a pH meter or bromothymol blue as external indicator. About 1 ml of NaOH shall suffice. The volume is made up with water. In the assay procedure one volume of basal medium stock solution is diluted to two volumes.

(ix) ***Agar medium for stock culture*** 5.0 g of Difco/Bacto or equivalent to yeast extract is dissolved in 200 ml water, 1 g each of anhydrous glucose and sodium acetate (or 1.7 g of Na $C_2H_3O_2.3H_2O$) + 3 g of agar is added, the mixture is heated on a steam bath with occasional stirring until the ingredients are dissolved. Only hot solution is filtered through cotton or cloth and added 10 ml to each of 20 culture tubes. The tubes are plugged with cotton and sterilized in the autoclave for 20 minutes at 120°C to 123°C. The tubes are cooled in an upright position and stored in the refrigerator.

(x) ***Culture medium for growing inoculum*** 5 ml of a solution containing 0.2 μg of niacin/ml is added to each of a series of culture tubes already containing 5.0 ml of (viii). The tubes are plugged with cotton and autoclaved at 120° to 123°C for 15 minutes. The tubes may be stored for several weeks, away from evaporation and contamination, etc.

(xi) ***Isotonic salt solution*** 900 mg NaCl is transferred into a 100 ml volumetric flask, diluted to volume with water, dissolving NaCl through shaking. 10 ml quantities of this solution are transferred to culture tubes, each cotton plugged then sterilized at 120-123°C for 20 minutes.

(xii) ***Bromothymol blue indicator*** 100 mg of this indicator is weighed in a beaker, 1.6 ml of 0.10 N NaOH is added contents

triturated with a stirring rod till dissolution. The contents are diluted with water to 250 ml. The solution is also made up by dissolving in a few ml of 95% ethanol, adding 1.5 ml of 0.1N NaOH and diluting to 250 ml with water.

Procedure

1. ***Preparation of stock culture*** Stab cultures are prepared in two or more agar stock culture tubes (ix above) using a pure culture. These are incubated for 20 ± 4 hours at 37.0° ± 0.5°C and stored in the refrigerator under aseptic conditions not longer than 6 days before transferring to new stab.
2. ***Inoculum preparation*** A day prior to use, cells are transferred from the stock culture to a sterile tube of inoculum culture medium (× above). This culture is incubated for 6-18 hours at 37°C. After securing the cotton plug with a rubber and/ adhesive tape/ pin, the tube is centrifuged, the supernatant is decanted, cells re-suspended.
3. ***Preparation of sample*** Sufficient sample material to contain 100 µg of niacin is weighed into a 250 ml flask. 100 ml of IN H_2SO_4 is mixed thoroughly into the above material, contents auto-claved at 120-123°C for 30 minutes, 1 N NaOH added to produce a pH of 6.8 using a pH meter or (xii) as an external indicator.

This solution is quantitatively transferred to a 1 litre volumetric flask, diluted to volume with water and contents mixed thoroughly. This solution is filtered through Whatman No.40 or equivalent filter paper. *This is the test solution* used in the preparation of assay tubes and niacin concentration should be 0.1 to 0.2 ug/ml to obtain valid results.

4. ***Preparation of standard tubes for titrometric method*** To duplicate tubes is added 0.0,0.5, 1.0,1.5, 2.0,2.5, 3.0 and 5.0 ml of (vii) (b), sufficient water to make the volume 5 ml + 5.0 ml of (viii).
5. ***Preparation of assay tubes*** To duplicate tubes is added 0.5,1.0, 2.0 and 3.0 ml aliquots of the test solution (3), sufficient water to make 5 ml in each tube + 5.0 ml of (viii).
6. ***Sterilization*** Contents of each tube are mixed thoroughly by rotating the tube vigorously in the palm of the hand. The tubes

are plugged with cotton or covered with caps and autoclaved at 120-123°C for 12 to 15 minutes.

7. ***Inoculation and incubation*** All tubes are cooled to the incubation temperature or below. Each tube is aseptically inoculated with 1 drop of inoculum and incubated for 72 hours at 37°C.
8. ***Titration*** Contents of each tube are transferred, successively to a 50 ml Erlenmeyer flask and rinsed with 10 ml H_2O. The contents of the flask are titrated with 0.1 N NaOH solution to a pH 6.8 using (xii) as an internal indicator.

Calculation

A standard curve is drawn for the assay by plotting the number of ml of 0.1 N NaOH used in titrating the standard tubes against 1 μg niacin/tube in the standard series. The niacin content of the tubes in the unknown series is determined by interpolation of the titre values on the standard curve, discarding values which are more than 0.4 or less than 0.04 μg of niacin/tube. The niacin content of each ml of test solution for each of the duplicate sets of tubes is calculated. The niacin content of the test material from the average of the values for 1 ml of test solution obtained from not less than three sets of these tubes (which do not vary by more than 10% from the average) is calculated using the formula.

μg Nicotinic acid (niacin) per g of food sample

$$\frac{\mu\text{g niacin/ml of solution} \times \text{volume}}{\text{weight of sample in g}} \times \text{dilution factor}$$

FOLIC ACID

Lactobacillus casei (ATCC No. 7469 or NCIM 2077) have specific requirements for folates, folic acid or monoglutamate, reduced free folates like tetrahydro-folic acid, THFA, 10-formyl THFA, 5, 10-methylene THFA, 5, 10-methenyl THFA, 5-methyl THFA, triglutamates and heptaglutamates, etc. for their growth. The growth response on a defined medium, complete in all respects, except the folic acid under test, is proportional to the medium up to a certain range. Either the acid or the turbidity produced by the organism is measured to determine the extent of growth and thereby the amount of folic acid in the test solution.

Estimation

Food stuffs usually contain the above forms of folates. For the estimation of heptaglutamate, the food extract is incubated with *folic acid conjugate,* the monoglutamate formed is determined by *L. casei.*

The organisms *Streptococcus faecalis (ATCC* No.8043) and *Pediococcus cerevisiae* (ATCC No.8081) may also be used in case of fractional estimation of different forms of folates.

The usual apparatus for microbiological assay of vitamins as described under riboflavin estimation is suitable for these assays also.

Reagents

(i) ***Casein hydrolysate*** 100 g of vitamin free casein or casein hydrolysate (acid digested as per IS: 7203-1973) is stirred with 250 ml of 95% ethanol for 15 minutes in an 800 ml beaker and filtered with suction. This step is repeated using another 250 ml of ethanol. This alcohol washed casein is transferred into a RB flask of at least 1 litre capacity, preferably one having two necks ground to standard taper. This casein is mixed well with 500 ml of constant boiling HCl. The flask is fitted with a glass stopper and a water-cooled condenser and refluxed over a low flame or on a hot plate for 10 ± 2 hours. 1:1 HCl is used for the hydrolysis. Careful and gradual heating avoids frothing during the initial stages of hydrolysis. Flask contents are mixed by shaking occasionally. A wet towel is kept ready to cool the flask should the reaction becomes too vigorous.

After refluxing, the flask is fitted with a condenser and receiving flask suitable for vacuum distillation and the maximum possible HCl is removed by concentrating the hydrolysate to a thick paste under reduced pressure. Air is introduced through a bleeder tube well into the bottom of the flask to minimise bumping during the final stages of the concentration. The distillation temperature here should be 75 ± 5°C. To get rapid and complete distillation at this low temperature, pressure is reduced considerably by a steam aspirator or a vacuum pump which must trap effectively HCl fumes.

The paste is redissolved in approximately 200 ml H_2O, the concentration step is repeated to remove additional amounts of HCl if a satisfactory hydrolysate has not been attained with a single concentration. The hydrolysate paste is dissolved in about 700 ml of water, pH adjusted to 3.5 with 40% NaOH. It is decolourised by stirring with 20 g of activated charcoal at room temperature to remove residual niacin and folic acid. Contents are stirred until a small test filtrate gets light straw colour to ensure removal of any niacin which might have remained in the alcohol-washed casein. It is filtered either through a large fluted filter paper or by suction as preferred.

pH of the filtrate is adjusted between 6 to 8 then diluted to 1 litre and stored under toluene and over $CHCl_3$ in the refrigerator. On standing, sometimes tyrosine precipitate is formed. As a good practice the solution is shaken, the suspended material as well as fluid portion is used. The insoluble material will dissolve when the entire medium is prepared.

(ii) ***Activated charcoal*** About 100 g of animal charcoal is weighed, 250 ml of HCl (1:1) is added, contents boiled for 2-3 hours. The mixture is diluted with hot H_2O, mixed well and filtered with a Büchner funnel under suction, the residue washed repeatedly with boiling H_2O until the filtrate is no longer acidic. Water is drained off, the residue dried in an oven at 110°C for 1 hour.

(iii) ***l-cystine*** 4.0 g of cystine is suspended in water, using conc. HCl to aid solution, the volume made up to 500 ml with water.

(iv) ***dl-Tryptophane*** 2.0 g of tryptophane is dissolved in water using conc. HCl to aid solution, pH adjusted to 3.5, volume made up to 500 ml, stirred with activated charcoal (ii) for 10 to 15 minutes, filtered and stored in the cold under toluene.

(v) ***Adenine-Guanine-Uracil (AGU) solution*** 200 mg each of adenine sulphate, guanine HCl and uracil in water using conc. HCl to aid solution, the volume is made up to 100 ml.

(vi) ***Xanthine*** 200 mg of xanthine is dissolved in water using NH_3 to aid solution, the volume made up to 100 ml with water.

(vii) ***Salt solution A*** 25.0 g each of $K_2\ HPO_4$ and $K\ H_2PO_4$ in 250 ml water.

(viii) ***Salt solution B*** 500 mg each of NaCl, $FeSO_4.7H_2O$ and $MnSO_4.4H_2O$ plus 10.0 g of $MgSO_4.7H_2O$ are dissolved, volume made up to 250 ml with water. A few drops of conc. HC1 are added to get a clear solution.

(ix) ***Pyridoxine and thiamine solution*** 50 mg pyridoxine or equivalent amount of pyridoxine HC1 and 200 mg of thiamine or equivalent amount of thiamine HC1 are dissolved in water and the volume made up to 100 ml with water.

(x) ***Biotin solution*** 250 mg of biotin in 500 ml of 50% ethanol in water.

(xi) ***Riboflavin solution*** 20.0 mg of riboflavin is made up to 100 ml, using acetic acid to help solution, may be heated on a water-bath if needed.

(xii) ***Nicotinic acid solution*** 50 mg in 500 ml of 50% ethanol in water.

(xiii) ***p-Amino benzoic acid (PABA)*** 12.5 mg in 100 ml water.

(xiv) ***Ca-Pentothenate*** 54.4 mg in 500 ml water.

(xv) ***Peptone solution*** 10 g of IS: 6853-1973 grade peptone is dissolved in 80 ml water, pH adjusted to 3.0 with HC1, the volume made up. The solution is stirred with 5 g of activated charcoal for 1 hour and filtered. This process is repeated twice with 2 g of charcoal.

(xvi) ***dl-Alanine*** 2% in water.

(xvii) ***Phosphate buffers (pH 7.2)*** 27.32 g of KH_2PO_4 and 5.60 g of K_2HPO_4 are dissolved in water and diluted to 1 litre with water.

(xviii) ***Ascorbic acid, 1%, charcoal treated*** 1 g ascorbic acid is dissolved in 20 ml water, 550 ± 50 mg of activated charcoal is added, the mixture is stirred gently for 15 minutes, filtered and to it is added an equal volume of (xvii). pH is adjusted to 6.1 and the solution is further diluted with an equal volume of water. This solution should be prepared fresh on the day of assay.

(xix) ***Folic acid standard solution***

(a) ***Stock standard*** 10.0 mg of folic acid is dissolved in 100 ml of 0.8% Na_2CO_3 solution to obtain 100 μg/ml. This solution may also be prepared in 0.01 N NaOH solution in 20% ethanol. This solution should be covered with toluene and stored in a coloured bottle under refrigeration.

(b) ***Working standard*** 0.1 ug/ml for assay with *L. casei.*

(c) ***Standard range*** *L. Casei:* 0 to 0.4 μg/ml.

(xx) Composition of basal medium for L. casei-for 100 ml

(i) 10 ml and (iv) 10 ml; (xv) 0.4 ml, (iii) and (vii) 5 ml each, (xvi), (viii), (v), (vi), (x) (1.5 ml of stock diluted to 100 ml), (ix) and (xi) 1 ml each; (xii) and (xiv) 2 ml each; (xiii) 0.3 ml, glucose 4 g and Na acetate. $3H_2O$-6.64 g.

pH is adjusted to 6.8, the volume made up to 100 ml, the solution is filtered.

(xxi) Na phosphate buffer 0.2 M-pH 6.1

(a) ***NaH_2PO_4 monobasic*** 31.2g/litre.

(b) ***Na_2HPO_4 diabasic*** 28.4 g/litre. 80 ml of (a) and 15 ml of (b) are mixed with 100 ml water.

Preparation of sample for the assay To 1.0 g of dry material or 10.0 g of fresh homogenised material is added 40 ml of 0.1 M (xvii) + 2.5 ml of 4% ascorbic acid solution (after treating with charcoal). The mixture is autoclaved for 15 minutes at 123°C then cooled to room temperature. pH is adjusted to 4.5, volume made up to 100 ml and the contents filtered. The filtrate is used for free folated estimation after suitable dilution. For the estimation of total folate, 5 ml of the filtrate is incubated, 4 ml of 1.2 M acetate buffer, pH 4.5, 1 ml of 100 M mercaptoethanol and 0.2 ml of human plasma or chicken pancreas or hog kidney and few drops of toluene are incubated overnight at 37°C. After incubation, the enzyme is inactivated by heating in a boiling-water bath for 3 minutes. Aliquots are used for the assay. A final concentration should be 0.1 ug/ml for *L. Casei.*

Preparation ofconjugase

(a) ***Chicken pancreas*** It is prepared by grinding the pancreas with five times its mass of distilled cold acetone. The mixture is left for overnight at the ambient temperature. The precipitate is filtered, washed with acetone and air dried powder is stored in the refrigerator. To 5/10 ml of filtrate is added 20 mg of

above powder, a paste is made with one drop of glycerine and in incubated overnight with a few drops of toluene.

An enzyme blank is run with similar incubation mixture using only the buffer and same amount of chicken pancreas.

(b) Hog Kidney

(i) Fresh hog kidney is blended with 3 ml of water/g of kidney in a Waring blender. The suspension is centrifuged the supernatant solution is filtered. It is frozen in 10 ml portions until ready for use.

(ii) ***Na-acetate buffer*** (1%) 5.0 g of anhydrous Na-acetate or 8 g of hydrated Na-acetate is dissolved in 400 ml water. pH is adjusted to 4.5 with acetic acid, and the volume made up to 500 ml.

(A) 0.5,1.0,2.0,3.0 ml are taken in duplicate, 1.0 ml of 0.25% ascorbic acid and sufficient water is added to bring the volume to 5.0 ml followed by 5.0 ml of basal medium.

Standard Levels 0.0, 0.5, 1.0, 2.0, 3.0 and 4.0 ml of (xix) (b) in duplicate is treated similar to (A) above.

Procedure

Preparation of standard tubes (L. casei) Concentrated to 0.1 µg/ ml, 0 to 5.0 ml of (xix) (b) in duplicate is dispensed and treated like (A) above. The tubes are covered with cotton and brown paper, autoclaved at 120-123°C for 12 minutes, then cooled to room temperature.

Preparation of assay tubes 0.0 to 0.3 ml of the test material is dispensed in duplicate, 1.0 ml of 0.25% ascorbic acid, etc. added and treated similarly as *L. casei* tubes above.

Inoculation and incubation Each tube is aseptically inoculated with one drop of inoculum except the blank tube. The tubes are incubated at 37°C for approximately 18 hours.

Turbidimetric method At the end of the incubation period, the assay tubes are removed and steamed for 5 minutes, cooled, turbidity in them is read in a colourimeter at 660 nm using the uninoculated blank tube to set the instrument at zero.

Calculation

A standard curve of the assay is drawn by plotting the optical density or turbidity reading on the X-axis against concentration of the vitamin on the Y-axis. The vitamin content of the tubes in the unknown series is determined by interpolation of the colourimeter readings on the standard curve. The average for 1 ml of test solution is calculated from the values obtained from not less than 3 tubes which do not vary by more than 10% on the average. The vitamin content of the test solution is calculated using the following relationship.

μgm of Folic acid/gm of sample

$$= \frac{\text{Average mg/ml} \times \text{Diluting factor}}{\text{Mass of the sample}}$$

Titrimetric method The contents of each tube are transferred to a 125 ml conical flask, the tubes are rinsed once with about 10 ml of H_2O, adding the rinsing to the flask. About 0.2 ml of 0.1% bromothymol blue is added and contents titrated with 0.1N NaOH to a green colour, pH about 6.8. A flask may be held as reference colour for about 15 ± 5 tubes only.

Calculation

A standard curve is drawn for the assay by plotting ml of 0.1 N NaOH used in tritration of the standard tubes against mmcg (mili micro centigram, 1 x 10^{10}g) of folic acid/tube in the standard series. Folic acid content of the tubes in the unknown series is determined by interpolations of the titre values on the standard curve. Values which vary more than 1.00 mmcg or less than 0.1 mmcg folic acid/ tube are discarded. Folic acid content of each of the duplicate sets of tubes is calculated so also of the test material from the average of the values for each ml of the test solution obtained from not less than 3 sets of these tubes which do not vary by more than 10% from the average, using the following formula:

mmcg/g of the sample

$$\frac{\text{Average mg/ml} \times \text{volume} \times \text{Diluting factor}}{\text{Mass of the sample}}$$

VITAMIN B_{12}

The microorganism *(Lactobacillus leichmannii)* has specific requirement for Vitamin B_{12} for its growth. The growth response on a defined medium complete in all respect, except vitamin B_{12} under test, is proportional to the concentration of the vitamin B_{12}, added to the medium up to a certain stage. The turbidity produced by the organism is measured to determine the extent of growth and thereby the amount of Vitamin B_{12}.

Reagents and media

(i) ***Casein hydrolysate*** 100 g of 'vitamin-free' casein hydrolysate conforming to IS: 7203-1973 is stirred with 250 ml of 95% ethanol for 15 minutes in an 800 ml beaker and filtered with suction. This step is repeated using another 250 ml of ethanol. The alcohol washed casein is transferred into a R.B. flask of at least 1 litre capacity, and preferably having two necks ground to standard taper. The contents are mixed well with 500 ml of constant boiling HC1. The flask is fitted with a glass stopper, a water-cooled condenser and then refluxed over a low flame or hot plate for 10 ± 2 hours. HC1 (1:1) is used for the hydrolysis. The flask is heated gradually to avoid frothing during the initial stages of hydrolysis. Flask contents are mixed by shaking. A wet towel is kept ready to cool the flask if the reaction becomes too vigorous.

After refluxing, the flask is fitted with a condenser and a receiving flask suitable for vacuum distillation. HCl is removed as much as possible by concentrating the hydrolysate to a thick paste under reduced pressure. Air is introduced through the bleeder tube, this tube is placed well into the bottom of the flask to minimize bumping during the final stages of the concentration. The distillation is carried out between 70 to 80°C. For pressure reduction either a steam aspirator or a vacuum pump should be used. Water-aspirator not suitable for this purpose. HC1 fumes must be carefully trapped while using vacuum pump.

The paste is redissolved in approximately 200 ml H_2O, concentration step repeated to remove additional amounts of HCl, if necessary. The acid concentration should be appreciably low so that subsequent neutralization must not yield salt which might retard bacterial growth on the basal medium.

The hydrolysate paste is dissolved in about 700 ml of H_2O, pH adjusted to 3.5 with 40% NaOH. The contents are decolourised with 20 g of activated charcoal at room temperature by stirring till a small test filtrate is of light straw colour. The decolourisation may be complete in 5 minutes or may require more than an hour depending upon the charcoal used. This step removes any niacin and residual vitamin B_{12} which may have remained in the alcohol-washed casein. Contents are filtered either through a large fluted filter paper or by suction.

The pH of the filtrate is adjusted to 6.8, this filtrate is diluted to one litre and stored under toluene and over $CHCl_3$ in a refrigerator. On standing, this solution may form a precipitate. This is mainly tyrosine. The solution is shaken and the suspended material is used so also the fluid portion. The insoluble material will dissolve when the entire medium is prepared.

(ii) ***Na-acetate buffer (0.1M)*** 6.8 g of Na-acetate and 2.9 ml of glacial acetic acid are together dissolved in water and the volume made up to 500 ml.

(iii) ***Asparagine solution*** 2.0 g of l-asparagine is dissolved in water to make 200 ml. It is stored under toluene in a refrigerator.

(iv) ***Adenine-Guanine-Uracil (AGU) solution*** 200 mg each of adenine sulphate, guanine HCl and uracil are dissolved by heating in 10 ml of dil. HCl (1:1), then cooled, the volurme made to 200 ml with water and this solution is stored under toluene in a refrigerator.

(v) ***Xanthine solution*** 200 mg xanthine is suspended in 35 ± 5 ml water heated to about 70°C, 6.0 ml NH_3 added, contents stirred till dissolution of solids. After cooling the volume is made up to 200 ml with water, this final solution is stored under toluene in a refrigerator.

(vi) ***Salt solution** A* 10 g each of K_2HPO_4 and KH_2PO_4 are dissolved in water and the volume made to 200 ml with water. After adding 2 drops of HCl, the solution is stored under toluene.

(vii) ***Salt solution B*** 4.0 g of $MgSO_4.7H_20$,200 mg each of NaCl, $FeSO_4.5H_2O$ and $MnSO_4.4H_2O$ are dissolved in water and preserved by adding 2 drops of HCl and stored under toluene.

(viii) ***Polysorbate 80 solution*** 20 g dissolved in 200 ml ethanol and stored in refrigerator.

(ix) ***Vitamin solution I*** 10 mg each of riboflavin and thiamine HC1,100 ug of biotin and 20 mg of nicotinic acid in sufficient quantity of 0.02 N acetic acid to make 400 ml. This is stored, protected from light, under toluene in a refrigerator.

(x) ***Vitamin solution II*** 20 mg of p-aminobenzoic acid, 10 mg of calcium pentothenate, 40 mg of pyridoxine HC1, 40 mg of pyridoxal HC1, 8 mg of pyridoxamine dihydrochloride and 2 mg of folic acid in sufficient quantity of neutralized (1:4) ethanol to make 400 ml. It is stored and protected from light, under refrigeration.

(xi) ***Tomato juice preparation*** Commercially canned tomato juice is centrifuged so that most of the pulp is removed. About 5 g/litre of analytical filter-aid is suspended in the supernatant liquid and filtered with the aid of reduced pressure through a layer of filter aid. If required refilteration is done until a clear straw-coloured filtrate is obtained; stored under toluene in a refrigerator.

(xii) ***Standard cyanocobalamin stock solution*** To a suitable quantity of cyanocobalamin reference standard (IP grade) or equivalent, accurately weighed, is added sufficient quantity of 25% ethanol to make a solution containing in each ml 1.0 μg of cyanocobalamin. It is stored in a refrigerater.

(xiii) ***Standard cyanocobalamin solution*** A suitable volume of (xii) is diluted with water to a measured volume such that after the incubation period as described under procedure, the difference in transmittance between the incubator blank and the 5.0 ml level of this standard (xiii) is not less than that which corresponds to a difference of 1.25 mg in dried cell mass. This concentration usually falls between 0.01 and 0.04 μg (milimicrogram), i.e. (10^{-9} g) per ml of (xiii). Fresh solution for each assay is imperative.

Media

(a) ***Basal medium, stock solution*** 100 mg 1-cystine and 50 mg 1-tryptophan are carefully dissolved in HC1 before adding to under-listed ingredients in sequence:

HC1 (IN) 10 ml, AGU (iv) 5 ml, (v) 5 ml, (ix) and (x) 10 ml (each). After this is added 5 ml each (vi), (vii) and (iii), (i) 25 ml, dextrose anhydrous 10 g, (ii) 5 mg, ascorbic acid 1 gm and (viii) 5 ml, in strict sequence.

(b) ***Culture medium*** 750 mg each of yeast extract, conforming to IS:7004-1973 peptone-conforming to IS: 6853-1973,1 g anhydrous dextrose, 200 mg of K_2HPO_4, all together are dissolved in 65 ± 5 ml H_2O. 10 ml (xi) + 1 ml of (viii) added. pH is adjusted at 6.8 with NaOH, H_2O added to make up to 100 ml. 10 ml portions are placed in test tubes, the tubes plugged with cotton, sterilized contents autoclaved at 121°C for 15 minutes The tubes are immediately cooled to avoid colour formation due to thermolysis.

(c) ***Suspension mediums*** A measured volume of (a) above is diluted with an equal volume of water. 10 ml portions of the diluted medium is placed in test tubes, sterilized and cooled as at (b) above.

Preparation of sample Assay sample must be representative and its vitamin content untampered. Powders and liquids should be mixed thoroughly till homogeneous. Dry materials like bread, biscuits, grains etc. should be ground and mixed thoroughly. Wet or fresh materials are minced with a knife/scissors or homogenised in a blender, if needed, in the presence of extracting solvent.

1. ***Extraction of Vitamin B_{12} from crude materials*** It is desirable to convert the more labile forms of Vitamin B_{12} to cyanocobalamin by cyanide treatment, therefore, B_{12} is extracted from crude materials with KCN solution (0.01%). It is effective (10 mg KCN for 10 μg of $Vitamin_{12}$ activity), pH 4.55 to 5.0 at 60°C for 30 minutes. Higher temperatures are applied to release bound forms of Vitamin B_{12} which has a maximum stability at pH 4 to 5; and acetate buffer pH 4.6 is recommended when higher temperature are used.

The absence of a response to Vitamin B_{12} may be due not only to bound forms but also to presence of inhibitors.

2. ***Assay preparation*** A suitable quantity of the assay material is accurately measured/weighed in an appropriate vessel containing for each gm/ml of sample taken, 25 ml of an aqueous extracting solution prepared just prior to use. It should contain, in each 100 ml, 1.29 g of Na_2HPO_4, 1.1 gm

of anhydrous citric acid and 1.0 gm of $Na_2S_2O_5$. The mixture is autoclaved at 121°C for 10 minutes. Insoluble particles of the sample extract are settled and filtration/ centrifugation is done, if needed. An aliquot of the clear solution is diluted with water so that the final test solution contains vitamin B_{12} activity approximately equivalent to that of the standard cy-anocobalamin solution added to the assay tubes.

Test Organism

(a) ***Stock culture of the Lactobacillus leichmannii:*** To 100 ml of culture medium is added 1.25 ± 0.25 g of agar conforming to IS:6850-1973 the mixture is heated on a steam bath until the agar dissolves. Approx. its 10 ml portions are placed in test tubes, suitably autoclaved at 121°C (exhaust line temperature) for 15 minutes. The tubes are cooled in an upright position. 3 or more tubes are inoculated by stab transfer of a pure culture of *Lactobacillus leichmannii,* ATCC No. 7830. Before first using a fresh culture in this assay, at least ten successive transfers of the culture are made in two weeks period, then incubated for 20 ± 4 hours at any selected temperature between 35 ± 5°C but held constant within ± 0.5°C, and finally stored in a refrigerator.

(i) Fresh stab cultures are prepared at least three time each week. More than 4 days cultures are unfit for preparing the inoculum. The activity of the micro organism can be increased by daily or twice-daily transfer of the stab culture, to the point where definite turbidity in the liquid inoculum can be observed 2 to 4 hours after inoculation. A slow-growing culture seldom gives a suitable response curve and may lead to erratic results.

(b) ***Inoculum*** Cells from the stock culture of *Lactobacillus leichmannii* are transferred to a sterile tube containing 10 ml of the culture medium. This culture is incubated for 6 to 24 hours at any selected temperature between 35 ± 5°C but held constant within ± 0.5°C. The culture is centrifuged, the supernatant liquid is decanted under aseptic conditions. Cells from the culture are suspended in 10 ml of sterile suspension medium. An aliquot is diluted with sterile suspension medium to give transmittance that corresponds to cell mass (as given at (a) above) of 0.50 to 0.75 mg/tube when read against the suspension medium set at 100% transmittance.

Procedure

Hard glass test tubes, about 120 x 150 mm in size are cleaned by suitable means, followed preferably by heating at 250°C for 2 hours because the test organism is highly sensitive to minute amounts of $vitamin_{12}$ activity and to traces of many cleaning agents.

1.0 or 1.5 ml, 2.0 ml, 3.0 ml, 4.0 ml and 5.0 ml of the standard cyanocobalamin solution is added in triplicate respectively to the test tubes. To each tube and to four similar tubes containing no standard cyanocobalamin solution is added 5.0 ml of basal medium stock solution and sufficient water to make 10 ml.

To similar test tubes is added in triplicate, 1.0 ml, 1.5 ml, 2.0 ml, 3.0 ml, 4.0 ml of the assay preparation respectively. 5.0 ml of basal medium stock solution and sufficient water is added to make 10 ml. One complete set of standard and assay tubes together is placed in one tube rack and the triplicate set is placed in a second rack or section of a rack, preferably in a random order.

The tubes are suitably covered to prevent bacterial contamination. The tubes and contents are autoclaved at 121°C for 15 minutes. This temperature must be attained within 10 minutes by preheating the autoclave, if required. The tube and contents are cooled as rapidly as practicable to avoid colour formation resulting from overheating of the medium. Uniformity of sterilizing and cooling conditions must be maintained throughout the assay, since packing the tubes too closely in the autoclave or overloading it may cause variation in the heating rate.

One drop of the inoculum is added aseptically to each tube so prepared except to two of the four containing no standard cyanocobalamin solution (two uninoculated blanks). The tubes are incubated at a temperature between 35 ± 5°C held constant within ± 0.5°C, until, following 20 ± 4 hours of incubation, there has been no substantial increase in turbidity in the tubes containing the highest level of standard during 2 hour period.

The contents of the tubes are mixed, one drop of a suitable antifoam agent solution may be added and transferred to an optical container, the transmittance of the tubes is determined. After agitating its contents, the container is placed in a spectrophotometer that has been set at a specific wave length between 540 and 660 nm, the transmittance is read when a steady state has been reached which is observed a few seconds after agitation when the reading remains constant for 30 seconds or more. The same time interval for the reading is adopted on each tube.

The transmittance is set at 1.00 for the uninoculated blank then the transmittance of the inoculated blank is recorded. If this transmittance reading corresponds to a dried-cell mass greater than 0.6 mg/tube or if there is evidence of contamination with a foreign microorganism, the assay results are discarded. Then with the transmittance set at 1.00 for the inoculated blank, the transmittance for each of the remaining tubes is read. If the difference between the transmittance observed at the highest level of the standard and that of the inoculated blank is less than the difference corresponding to a dried-cell mass of 1.25 mg/tube then the results of assay are discarded.

Identification of Vitamin B_{12}

The absorbency of the solution prepared for the assay in a 1 cm. quartz cell with a suitable spectrophotometer is determined, using water as the blank. Maxima within ± 1 nm are found at 278 and 361 nm and within ± 4 nm at 548 nm. The ratios of absorbencies are,

(a) d_{361}/d_{278} is not less than 1.62 and not more than 1.88,

(b) d_{361}/d_{548} is not less than 2.83 and not more than 3.45.

Assay On a microbalance about 2 mg of Vitamin B_{12} is accurately weighed and transferred to a 50 ml volumetric flask with the aid of 15 or 20 ml water and the volume is exactly made within water and contents mixed well. The absorbency of the solution is determined in a 1 cm-quartz cell at 361 nm with a suitable spectrophotometer using water and blank. The percentage purity of Vitamin B_{12} is calculated by the formula,

$$\frac{d361}{0.0207} \times \frac{1}{\text{sample mass (mg/10ml)}} \times \frac{100}{100 - \text{loss or drying}}$$

The final stock solution for microbiological assays is prepared to contain, for example, 10 mg of pure vitamin B_{12} and 10 mg of KCN/ml in 0.1 M acetate buffer of pH 4.6. The solution is dispensed in 1 ml ampoules which are sealed and sterilized at 100°C for 20 minutes and stored in a refrigerator for not more than *one year.* A new ampoule should be used for each assay.

Repeatibility The repeatability of the results should be within the range of ± 5%.

DETERMINATION OF PYRIDOXINE (VITAMIN B_6)

1. *Reagents*

(i) Stock solutions

(a) The following are dissolved in water and the volume made up to 100 ml. KH_2PO_4 1.! g; KC1 0.85 g; $CaCl_2$, 0.25 g; $MgSO_4$ 0.25g; $MnSO_4$ and $FeCl_3$ 5 mg each. Any precipitation is brought into solution by adding a few drops of HC1.

(b) ***Casein hydrolysate*** 100 g of vitamin free casein or its hydrolysate is stirred with 250 ml of 95% alcohol for 15 minutes in a 800 ml beaker and filtered through suction. This is repeated with a fresh 250 ml portion of alcohol. Thus washed casein is transferred into a round bottomed flask of capacity not less than 1 litre, and preferably one having two necks grounded to standard taper. It is mixed with 500 ml of constant boiling HC1. The flask is fitted with a glass stopper and a water-cooled condenser, contents refluxed over low heat for 10 ± 2 hour. HC1 (1:1) solution is used for the hydrolysis and reactants heated carefully and gradually. This precludes frothing likely to appear at the initial stages of these hydrolysis. The flask contents are mixed occasionally by shaking. A wet towel is kept ready to cool the reactants in the event of the reaction getting vigorous.

The flask is fitted with a condenser only after the refluxing part is over and with a receiving flask suitable for vacuum distillation. While the hydrolysate is concentrated, almost entire HC1 gets removed under reduced pressure. Air is introduced in the flask to effectively reduce bumping during the final stages of concentration through a bleeder tube placed well into the bottom of the flask. The distillation temperature should be 75 ± 5°C.

The distillation is completed rapidly at this low temperature by pressure reduction by a steam aspirator or a vacuum pump. Water aspirators are of little help. Effective trapping of HC1 vapours is essential, especially under vacuum operation.

The concentrated hydrolysate becomes pasty, it is redissolved in about 200 ml water, the process of concentration is again done to remove more of HC1, if necessary. The acid concentration should be low enough lest due to subsequent neutralization may yield salt which may retard bacterial growth on the basal medium.

The hydrolysate paste is redissolved in about 700 ml of water, pH 3.5 is adjusted with 40% NaOH. Activated charcoal 20 g at room temperature, is employed to decolourise the solution. The contents are stirred till a small test filtrate is of light straw colour. This decolourisation may be complete in 5 minutes or may take more than an hour depending upon the quality of charcoal employed for decolourisation.. This step removes any left over niacin in the alcohol washed casein. A quick filteration is effected either through a large fluted filter paper or by suction.

The pH of the filtrate is adjusted to 6.8, the filtrate is diluted to 1 litre. This is stored under toluene and over $CHC1_3$ in a refrigerator. This solution on standing may form a precipitate consisting mainly of tyrosine. As a good practice the solution should be shaken; the suspension and the fluid portion used. The insoluble material will dissolve during the preparation of entire medium.

(c) ***Biotin solution*** 4 mg of biotin is dissolved in 100 ml water by warming if required and further dilution 1:10 is made with water. This is prepared afresh.

(d) ***Ca-pantothenate solution*** 5 mg of Ca-D-pantothenate, 50 mg of meso-inositol and 5 mg of niacin, together are dissolved by warming in 100 ml water.

(e) ***Thiamine solution*** 5 mg thiamine HC1 in 25 ml water.

(f) ***K citrate buffer*** 10 g K citrate + 2 g citric acid dissolved in 100 ml water.

(ii) Basal medium

Stock solution (i) (a)	10 ml
Stock solution (i) (b)	8 ml

Stock solution (i) (c)	0.4 ml
Stock solution (i) (d)	10 ml
Stock solution (i) (e)	0.25ml
Stock solution (i) (f)	10 ml

All these are mixed, pH adjusted to 5.5, volume made up to 100 ml, then 60 ml of glucose solution (100 g in 600 ml water) is added.

(iii) Medium for stock culture

Malt extract (IS: 7591–1975)	0.3 g	All of micro-biological grade
Glucose	1.0 g	
Yeast extract (IS: 7004–1973)	0.3 g	
Peptone (IS: 6853–1973)		
Agar (IS: 6850–1973)	2.1 g	

All the above ingredients are dissolved by heating in 100 ml H_2O. 8 ml portions are dispensed into test tubes while the solution is hot. The test tubes are plugged with cotton and autoclaved at 121-123°C for 15 minutes. The tubes are placed in a slanting position and cooled. *Saccharomyces carlsbergensis* is sub-cultured every fortnight in slope cultures.

Procedure

(iv) ***Inoculum*** Freshly grown agar slope is washed, after 20 hours then cultured with 10 ml saline and transferred under saline conditions into a sterile tube. This solution is very rich in organisms, therefore a suitably diluted inoculum of 15% turbidity is prepared and used in inoculating the standard and samples. This 15% turbidity or 85% transmission corresponds to about 0.071 optical density.

(v) ***Stock standard*** 50 mg of pyridoxine is dissolved in 100 ml water and stored under refrigeration.

(vi) ***Working standards*** 1 ml of (v) above is diluted to 10 ml. 1 ml contains 5 μg. This is further diluted hundred fold. 1 ml = 50 μg. Standards containing 0, 5, 10, 15, 20, 25,30, 40 and 50 μg of pyridoxine are dispensed into corresponding 100 ml volumetric flasks, the volumes made up with sterilized water. These standards should be prepared daily.

Preparation of samples It is imperative that the assay sample is representative of the whole, with no damage to the pyridoxine content. Powders and liquids must be homogeneous. 10 g of dry materials like bread, biscuits and grains are powdered and mixed thoroughly. Wet or fresh materials must be minced or homogenised in the presence of extracting solvent in a blender.

2 g of fresh material/dry powder is suspended in 20 ml H_20,20 ml of 0.5N H_2SO_4 is added, contents autoclaved for 4 hours at 120-123°C, cooled, pH adjusted to 5.5 the volume made up to 100 ml with water, mixed well and filtered. Suitable aliquots from the filtrate in triplicate are taken at 3 levels for pyridoxine assay. In the case of flesh foods, 10% aqueous homogenates are prepared in a Waring blender. 1 ml is diluted to 5 ml with H_2O, 5 ml of IN HC1 added, contents autoclaved for 4 hours at 120-123°C, cooled, pH adjusted to 5.5, volume made up to 50 ml and finally filtered. Suitable aliquots in triplicate at 3 levels are taken for pyridoxine assay.

The method is based on the observation that *Saccharomyces carlsbergenesis* requires specified vitamins for growth using a basal medium complete in all respects expect for pyridoxine and growth responses of the organism are comparable quantitatively in standard and unknown solutions.

The standards or samples are dispensed into 5 ml Corning conical flasks. The volume is initially made up to 1.0 ml. 8 ml of (ii) is added and sterilized for 12 minutes at 120-123°C. 1 ml of (iv) is added under sterile conditions and incubated at 37°C for 20 hours. The turbidity is measured in a colourimeter at 660 nm using the uninoculated blank tube to set the instrument at zero.

Calculation

A standard curve is drawn for the assay by plotting optical density of turbidity reading on the X-axis against concentration of the vitamin on the Y-axis. The vitamin content of the tubes in the unknown series is determined by interpolation of the colourimeter reading on the standard curve. The average for 1 ml of test solution is calculated from values obtained from not less than 3 tubes which agree within 10% on the average.

Pyridoxine (Vit B_6) mg/g of sample

$$= \frac{\text{Average mg/ml} \times \text{Diluting factor}}{\text{Mass of the sample}}$$

The results should be reproducible within the average of ± 5%.

VITAMIN C

Titration method for ascorbic acid estimation In this titration the dye 2, 6-dichlorophenol indophenol (DCP) is used as an indicator which is reduced by the ascorbic acid to a colourless form.

Reagents

(a) For fruits and Vegetables

(i) ***DCP dye solution*** 800 mg of the dye is dissolved in about 500 ml hot, previously boiled and cooled water. If needed the contents are filtered and the volume made up to 1 litre with water. Shelf life of this solution when kept cool in a dark bottle is 7 days.

(ii) ***MPA solution*** 15 g of m-phosphoric acid (HPO_3) is dissolved in 40 ml glacial acetic acid and 200 ml H_2O. The contents are diluted to 500 ml and filtered. Under refrigeration its shelf life is also less than 7 days.

(iii) ***Standard Ascorbic acid (AA) solution*** 200 mg L-ascorbic acid is dissolved in 10 ml MPA solution and the volume made up to 100 ml with water. 1 ml of this solution is further diluted to 100 ml with water and stored under refrigeration.

This is solution A whose 1 ml contains 2 mg ascorbic acid/100 ml.

(b) For Milk

(t) ***DCP dye solution*** 125 mg of the dye is dissolved in warm water, filtered and the volume made up to 50 ml with water. 5 ml of this solution is further diluted to 100 ml with water. It should be used within 7 days.

(ii) ***MPA solution*** 25 g of HPO_3 sticks are dissolved in water and the volume made up to 500 ml. It is filtered, and stored under refrigeration and used within 7 days.

(iii) ***Standard Ascorbic acid (AA) solution*** 100 mg of L-ascorbic acid is dissolved in 5% (m/v) HPO_3 solution, diluted to 500 ml in a volumetric flask using more HPO_3 solution.

Procedure

A. Fruits and Vegetables

(a) ***Standardisation of reagents*** The dye solution is standardised by pipetting 5 ml of (a) (iii) into a boiling tube and titrating rapidly with the dye solution. The blue dye is first decolourised by the ascorbic acid, but the dye is run continually, with shaking, until faint pink colour persists for 15 seconds. Number of mg of ascorbic acid that are equivalent to 1 ml of dye solution is calculated. Since ascorbic acid solutions are unstable, this standardisation needs to be done daily.

(b) ***Determination of AA in fruit juice*** 10 ml of the fruit juice is pipetted into a conical flask, 2 ml of (a) (ii) is added. Concentrated juice is diluted by pipetting 20 ml into 100 ml volumetric flask and the volume made up with (a) (ii). 10.00 ml is titrated against the dye at (a).

(c) ***AA Determination in fruits and Vegetables*** 15 ± 5 g of the fruit/ vegetable depending on the amount of AA likely to present is weighed and is either ground in a pestle and mortar with a little clean sand and 15 ± 5 ml MPA mixture. This mixture reduces oxidation of the ascorbic acid, inactivates enzymes and reduces interference from any iron present. Contents are further ground with 50 ml more of the MPA mixture and strained through muslin. The pestle and mortar is washed with more of the mixture, this too strained as before. The water wash is also sieved, the extract volume is made up to 250 ml with MPA mixture.

Alternatively, a macerator or liquidiser can be used instead of the pestle and mortar, the extract is centrifuged. The residue is twice washed with MPA mixture after pouring off the supernatant liquid. The two supernatants are combined and the volume made up to 250 ml as above.

10 ml portions are titrated with the dye and AA content is calculated.

Calculation

The dye solution is standardised. If A ml of dye solution is consumed to react with 5 ml of ascorbic acid and B ml is consumed to react with 10 ml of fruit juice, then

1 ml dye solution = I/A mg of AA and

100 ml fruit juice = B x 10/A mg of AA.

Note: Dilutions if any must be considered for calculations.

D. Milk

(a) ***Standardisation of DCP solution*** 50 ml of AA solution (iii) is pipetted into a 100 ml volumetric flask and the volume made up with MPA solution, 2 aliquots of 25 ml are pipetted into each of 2 conical flasks. These are titrated with DCP dye solution as before.

(b) ***Estimation of AA*** 50.00 ml milk is diluted with 50.00 ml of MPA solution, contents filtered, 25 ml filtrate aliquots are titrated against DCP solution until the dye colour is just discharged.

Calculation

If the average volume of dye required for standardisation against 25.00 ml of AA solution V = T, and average dye required for titration against 25.00 ml of milk filtrate = M, then 100 ml of original milk contained.

$(M \times 0.25 \times 2 \times 4)$ T = 2MT mg of ascorbic acid.

COLOURIMETRIC ESTIMATION OF VITAMIN D

$SbCl_3$ reacts with Vitamin D producing colour whose measurement leads to quantitative determination of the vitamin.

Reagents

(i) ***Cotton seed oil*** Suitability of this oil is examined by saponifying 10 g of oil, unsaponifiable residue (USR) is dissolved in 10 ml of solvent hexane. 0.4 ml of $FeCl_3$ solution (1:1000) is taken in a separate container, 12 ml of, 1:6000 solution of α, α-dipyridyl in absolute alcohol is mixed with $FeCl_3$ solution, after 5 minutes the absorbance of this mixture is read in a 1 cm cell at 520 mμ, with a suitable spectrophotometer using

absolute alcohol as the blank. 0.2 ml of solvent hexane solution of USR is added and after 5 minutes its absorbance is also noted. The difference between these two readings must not be less than 0.125.

(ii) ***Solvent Hexane*** When measured in a 1 cm cell at 300 mμ with a suitable spectrophotometer, against air as the blank, the absorbance shall not be more than 0.070, otherwise it is redistilled.

(iii) ***KOH*** 780 g dissolved in litre.

(iv) ***Ethylene Dichloride, EDC*** It is purified by passing through a column of granular chromatographic grade silica gel.

(v) ***Colour reagent*** Preparation of the colour reagent includes preparing.

  (a) ***Solution A*** The entire contents of a previously unopened 113 g bottle of dry, crystalline $SbCl_3$ are emptied into a flask containing about 400 ml of EDC, about 2 g of anhydrous alumina mixed, contents filtered through filter paper into a clear glass, glass-stoppered container calibrated at 500 ml. EDC is added to make up the volume to 500 ml and contents mixed, the absorbance of the solution, measured in a 20 mm cell at 500 mu with a suitable spectrophotometer, against EDC should not be more than 0.070.

  (b) ***Solution B*** 100 ml of acetylchloride + 400 ml of EDC are mixed under a hood.

  (c) ***Solution A : B :: 9:1*** are mixed to obtain the colour reagent, which is stored in a tight container. This solution should be used within 7 days. However if the solution develops into a coloured one, then it is rejected outright.

Apparatus

A. Chromatographic tubes

(i) ***First column*** For descending chromatography (DC) a tube 2.5 cms (i.d.), about 25 cms long and constricted to 8 mm dia for a distance of 5 cms at the lower end is arranged by inserting at the constriction point sintered-glass disc of coarse-porosity/a small plug of glass-wool. The constricted portion may be fitted with an inert plastic stop cork.

(ii) Second Column There are three sections of this column.

1. A flared top section, 18 mm (i.d.) x 14 cms,
2. A middle section, 6 mm (i.d.) x 25 cms and
3. A tapered constricted lower exit-tube, approximately 5 cms long. A small plug of glass-wool is inserted in the upper 1 cm portion of the constricted section.

B. Chromatographic columns

(a) ***1st column*** Screw capped, wide mouth bottle containing 125 ml of isooctane and 25 g of chromatographic grade siliceous earth are shaken for slurry formation. 10 ml of polyethylene glycol 600 is added dropwise and mixed vigorously. The bottle cover is replaced and contents vigorously shaken for 2 minutes. 50% of the resulting slurry is poured into chromatographic tube and let to settle by gravitation. Gentle suction helps in slurry setting. The remainder slurry is also poured in small portions, packing each new portion with a 20 mm disc plunger. After the formation of a solid surface, vacuum is removed, 2 ml of isooctane is added.

(b) ***Second column*** The middle section of the tube is packed with 3 gm of moderately coarse chromatographic grade fuller's or equivalent earth through about 125 mm Hg suction. This earth should loose water content 8.75 ± 0.25% on drying.

(c) ***Standard preparation*** 25 mg of USP or equivalent grade Ergocalciferol Reference Standard (ERS) accurately weighed is dissolved in sufficient isooctane to yield a concentration of about 250 μg/ml. It is kept under refrigeration.

On assay day, 1 ml of the standard solution is pipetted into a 50 ml volumetric flask, it is desolventised with the help of a stream of nitrogen, the residue is dissolved in EDC and the volume finally made with EDC itself.

C. Sample preparation

The assay sample is accurately weighed/measured.equivalent to not less than 25 μg, preferably about 250 μg of ergocalciferol. This sample should have about 3000 USP units of Vitamin A, *perse,* otherwise an equivalent 1.5 mg of Vitamin A Acetate is added to provide the needed pilot bands in the subsequent chromatography. 2 ml of cotton seed oil is added followed by a volume of KOH solution (iii) above, corresponding to 2.5 ml of each gram of total weight of the sample + cotton seed oil, but more than 15 ml; then is added 50 ml of alcohol. Contents are refluxed vigorously on a steam bath for 20 minutes, for 30 minutes for samples weighing more than 5 g. Absence of oil globules connotes complete saponification. To samples not completely saponified, 5 ml of KOH (but never more than a total of 100 ml) is added and contents reheated, cooled and transferred to a 500 separating funnel, rinsing the saponification flask with a total of 50 ml of H_2O, adding each rinse to the separating funnel. About 4.5 g of anhydrous Na_2SO_4 and 150 ml of solvent hexane is added, contents shaken vigorously for 2 minutes, layers let to separate, aqueous layer discarded, hexane layer extracted twice with 50 ml portions of solvent hexane, extracts combined after discarding the aqueous layer of each extraction.

The combined hexane extracts are washed with 50 ml portions of water until the last portion is alkali free, checked with phenolpthalein. After 5 minutes, the separated water, if any, is discarded, organic phase is transferred into a 300 ml tall-form beaker containing about 5 g of anhydrous Na_2SO_4. The contents are stirred for 2 minutes, the supernatant solution is decanted into a 500 ml, tall-form beaker. Na_2SO_4 is rinsed with 4 x 25 ml portions of solvent hexane, combining the rinsings with the original extract. The total volume is brought down to about 300 ml by evaporation on a steam bath, the concentrate is transferred to a small round bottom evaporation flask. The beaker is rinsed with 4 x 5 ml portions of solvent hexane, adding the rinsing to the flask. Through vacuum, in a water bath maintained below 40°C, alternatively with a N_2 stream at room temperature, the contents are completely desolventised. The residue is again dissolved in solvent hexane, the solution transferred to a 10 ml volumetric flask, volume made up with solvent hexane.

D. Procedure

As a precaution, vitamin solutions should always be protected from oxygen and actinic light. Actinic light relates to sun rays capable of causing photochemical changes.

1. First column chromatography Just as the 2 ml iso-octane vanishes into the surface of the prepared first column, 2 ml of the sample preparation is pipetted on to the column. The moment this sample preparation reaches the column surface the first of 3 2-ml-portions of solvent hexane is added, followed by the other two but only when the former portion disappears in the column. Solvent hexane in 5 to 10 ml portions is added continually up to 100 ml. If needed, flow rate is adjusted between 3 to 6 ml/minute by application of gentle pressure at the top of the chromatographic tube.

The first 20 ml of the effluent is discarded, the rest collected. The column is examined under ultra violet light at intervals during the chromatography. The flow is stopped when the front of the fluorescent representing vitamin A is about 5 mm from the bottom of the column.

The ultra violet lamp should provide weak radiation in the 300 mm region. It is frequently essential to use a narrow aperture/ screen with commercial lamps to reduce the amount of radiation to the minimum required to visualise the Vitamin A on the column.

The eluate is transferred to a suitable evaporation flask, solvent hexane removed completely under vacuum at a temperature not above 40°C or with N_2 stream at room temperature. The residue is dissolved in about 10 ml of solvent hexane.

2. Second column chromatography The above obtained solvent hexane solution is added on to the second column like done on the first column earlier, the evaporation flask is rinsed with a total of 120 ml of solvent hexane in small portions, adding each portion to the second column, let to flow through the column, the effluence discarded when about 12 ml of the solvent hexane is left on the column surface, 75 ml of benzene is added and eluted with gentle suction of about 125 mm Hg, the eluate is collected. Benzene is also evaporated under vacuum below 40°C or with N_2 stream at room temperature.

3. Assay preparation The second column residue is dissolved in a small amount of EDC (iv), solution transferred to a 10 ml volumetric flask, the volume made up with EDC (iv) to obtain assay preparation.

4. Colour development 1 ml of assay preparation is pipetted into serially numbered 1,2 and 3 colourimeter tubes of about 20 mm i.d. Into tube 1, 1 ml of the standard solution (B) (c); into tube 2, 1 ml of EDC (iv) and into tube 3, 1 ml of a mixture of equal volumes of acetic anhydride + EDC (iv) is added. To each tube is added quickly, and preferably from an automatic pipette, 5.0 ml of colour reagent (v) (c), contents mixed. After 45 seconds, absorbances of the three solutions are measured at 500 mμ with a suitable spectrophotometer, using EDC (iv) as blank. After 45 seconds of taking the first reading of each solution, another reading of absorbances of the solutions in tube 2 and 3 at 550 mn are taken similarly. Denoted corresponding readings are A^1_{500}, A^2_{500}, A^3_{500}, A^5_{550}, and A^3_{550}.

Calculation

μgm of Vitamin D in the portion of the sample is = Cs/C × (Aμ/As), where,

Cs = Concentration of Vitamin D (μg/ml) of the standard preparation,

C = Concentration of sample as g in each ml of the final solution,

Aμ = Value of $(A^2_{500} - A^3_{500} - 0.67\ (A^2_{550} - A^3_{550})$ as above,

As = Value of A^1_{soo} - A^2_{soo} .

ESTIMATION OF VITAMIN E

Vitamin E

The determination of Vitamin E is carried out by colourimetry. Vitamin E is oxidised by $FeCl_3$ and the $FeCl_2$ so formed gives, on reaction with α, α–dipyridal, a red colour which is measured at 515 nm. The procedure allows separation and estimation of α- and β-tocopherol from other tocopherols, and can also determine the total tocopherol content.

The principle of this estimation is that the extract containing total tocopherol is purified from interfering substances, such as vitamin A and carotenoids by reduction with zinc and treatment with $SbCl_3$, and fats are eliminated by saponification. α and β-tocopherols are separated together from the other tocopherols by chromatography on alumina and determined colourimetrically with iron-dipyridal. Calculations are carried out with the aid of a calibration curve and a recovery test.

Total tocopherol The α–, β–, γ–, δ, and ε-tocopherols still adsorbed on the alumina are consequently eluted together and determined colourimetrically. The sum of the results obtained by both assays give the total tocopherol content. Figure 9.1 shows a chromatographic column.

Reagents

(i) ***Elution Mixtures*** It needs following rectified solvents

(a) ***Methanol*** 10 litres of methanol distilled over 5 g of $KMnO_4$+ 10 g of KOH.

(b) ***Petroleum ether (PE)*** Boiling points 30-40°C or 40-60°C or 90-115°C distilled over conc. H_2SO_4.

(c) ***Ethanol*** Absolute, purified like methanol above and distilled repeatedly until no fluorescence can be detected in U.V. light.

(d) ***Diethyl ether*** Free from per oxide, treated with $FeSO_4$.

(e) *Mixture no. 1*; b : d :: 1 : 1.

Mixture no. 2; Cyclohexane : d :: 99 : 1

Mixture no. 3; b:d :: 85 : 15

(ii) ***Al_2O_3 for chromatography*** Al_2O_3 is activated by heating at 400°C for 3 ± 1 hours, cooled under vacuum. 100 g of this powder is shaken with 10-15 ml water till all clots disappear

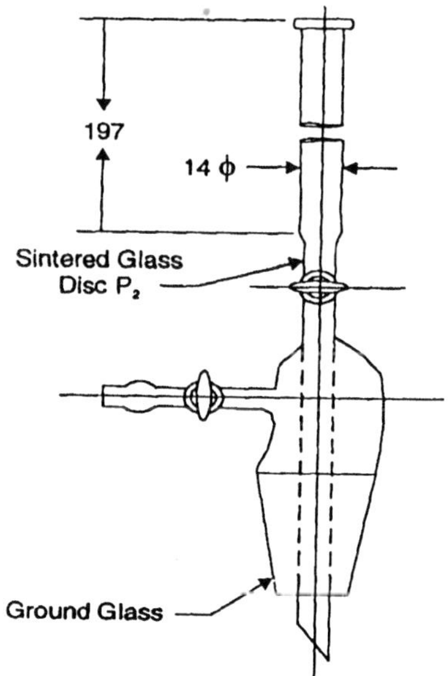

All dimensions in millimetres

Chromatographic Column

Figure 9.1

and the powder flows freely. The deactivated alumina is stable for one day and needs to be prepared one hour before use.

Note: It is essential to standardize the activity of each lot of alumina. The test is carried out with vitamin A which is added to the column and the column is developed with benzene or hexane. Vitamin A is located by $SbCl_3$ reagent under U.V. light. Suitability of alumina should be checked with pure α-tocopherol and with- β- or γ-tocopherols.

(iii) ***$FeCl_3$ solution (0.2% alcoholic)*** 200 mg of $FeCl_3$ is dissolved in 100 ml of (i) (c). The solution is sensitive to light hence prepared shortly before use and stored in brown coloured bottle and kept away from light.

(iv) ***α, α-Dipyridyl*** 0.5%, alcoholic solution.

(v) ***$SbCl_3$ solution*** 22 g of $SbCl_3$ are added to 100 ml $CHCl_3$ and refluxed till complete dissolution. When cold, the reagent is kept in a brown bottle over highly activated Al_2O_3. Before use it is agitated and filtered.

(vi) ***HCl*** (a) conc., 38%; (b) Dil, IN approx. 94 ml of (a) in 1 litre of water.

(vii) ***Water*** The distilled water for washing the diethyl ether-PE extracts should be saturated with a mixture of equal parts of the two solvents.

(viii) ***Vitamin E test solution*** About 95 mg of pure α, α-tocopherol is accurately but rapidly weighed in a 100 ml volumetric flask, dissolved immediately in (i) (c), the volume made up with (i) (c). The solution can be kept up to 10 days at 4°C in dark. It is diluted ten times with (i) (c) before use.

A. Extraction of Vitamin E from Water Dispersed Preparations

Rose-Gottleib procedure is followed, according to which 40 ml of the liquid preparation is mixed in a 250 ml glass-stoppered cylinder with the following reagents and shaken thoroughly after each addition.

10 ml of 10% NH_3 solution, 40 ml of (i) (c)/60ml each of mixture No. 1 The mixture is let to stand for at least 15 minutes in the dark. When complete separation of layers is achieved, the total volume

of the ether-PE solution is determined, and an aliquot thereof is evaporated carefully, and the residue further treated as described.

Procedure

(i) ***Calibration curve*** Increasing amounts 0-200 μg of Vitamin E of (viii) above each measured into a 10 ml volumetric flask and diluted with ethanol up to a volume of 8 ml. 1 ml each of (iv) and (iii) are added consecutively to the flask and the stop watch is started. The colourimeter is set to the zero point with (i) (c) and after two minutes, the absorption of the Vitamin E solution is measured at 515 nm.

Note: It is imperative that the calibration curve runs concurrently with the sample. The procedure should be carried out in non-actinic light.

B. Extraction of Tocopherols from Dry Products with Low Lipid Content

10 g of the sample is accurately weighed, for larger samples the quantity of the solvents should be increased accordingly, containing at least 1 mg of Vitamin E, in a non-actinic glass vessel. After addition of about 100 mg each of pure, sodium ascorbate powder, *Lipase, Clarase 900, Pig* pancreatic powder, the sample is stirred into a paste with 35 ml of warm (70°C) water, the flask let to stand for 20 minutes on a water-bath at 45°C in the dark. The mixture is agitated with 6.5 ml of 10% NH_4OH and 35 ml (i) (c) then cooled to room temperature and shaken thoroughly with 70 ml of (i) (d) followed by gentle agitation with 70 ml of (i) (b) and kept aside until the two layers get separated. The alcohol in the separated petroleum ether layer is removed by washing approximately 100 ml of PE with 2 x 10 ml of water. An aliquot of the washed extract containing at least 200 ug of total tocopherol is measured into each of the two RB flasks. To one of the flasks is added an amount of standard Vitamin E solution (viii) above corresponding to its expected α-tocopherol content.

C. Extraction of Tocopherols from Dry Products Containing Lipids

The procedure is same as at (B) above, however, the extract is evaporated in the recovery flask in vacuum or under nitrogen and the samples are saponified.

D. Extraction of Tocopherols from Fats and Oils

8 ± 2 g of the test sample, butter, etc. is weighed *accurately* in a 100 ml volumetric flask then dissolved in (i) (b) and the volume made up with (i) (b) only. A measured aliquot of this solution corresponding to a total tocopherol content of at least 100 mg and to a fat amount not exceeding 3 g; if a larger amount of fat is needed, the quantity of reagents and solvents necessary for further treatment are also increased accordingly; is measured twice into two separate RB flasks and treated with 100 mg of pure sodium ascorbate powder. In one flask (Recovery Test) an amount of Vitamin E test solution (viii) corresponding to the α-tocopherol content of the assay solution is added. Subsequently, both normal and recovery tests are treated similarly. The solvent is evaporated in vacuum, the residue is saponified as described after the following note.

Note: Peroxides eventually present in fats may destroy Vitamin E during the assay and should be removed by taking up the residue in 10 ml of elution mixture No. 3, filtering it through a 5 cm high column of alumina (ii) above, The total tocopherol is washed out of the column with 80 ml of elution mixture No. 3. The residue obtained after evaporation of the filtrate in vacuum is saponified and treated as at (D) above.

Saponification 15 ml of approx. IN methanolic KOH, prepared shortly before use, is added to the residue obtained from (C) and (D) followed by 150 ± 50 mg of sodium ascorbate powder, contents refluxed for 15 minutes in a water-bath at 65 ± 5°C. The mixture is stirred from time to time, a magnetic stirrer is suitable.

Extraction of unsaponifiable matter (USM) The soap solution is cooled to approximately 40°C and transferred quantitatively into a 250 ml glass-stoppered cylinder with 50 ml of methanol (i) (a), then cooled to room temperature and extracted by thoroughly shaking it with 100 ml of PE (i) (b). As soon as the layers have separated, 10 ml of water is added, shaking is repeated to achieve quantitative separation of methanol-PE mixture. The aqueous layer is sucked out, the residual extract is washed with 20 ml of H_2O, followed by 20 ml HCl (vi) (b).

Purification of the extract The extract obtained directly from dry products containing low amounts of lipids (B) above is evaporated in vacuum/under nitrogen. The residue is mixed thoroughly with 3 ml of $SbCl_3$ (v). 10 ml of (i) (c) is added after 5 minutes followed by the addition of 10 ml of (vi) (a) and about 50 mg of zinc powder. Zinc powder is again added twice or thrice until the solution becomes colourless. During this operation the flask is kept cooled to prevent over heating of the mixture above 40°C. After this step the mixture is quantitatively transferred to a 250 ml glass stoppered cylinder with 30 ml of (i) (c) and shaken thoroughly with 540 ml of (i) (d), then agitated with 50 ml of PE (i) (b) and, after the layers have separated, with 50 ml water. The impure products formed after reaction with $SbCl_3$ and some zinc powder contained in the extract must be removed before the assay is further proceeded. Consequently after the layers have separated an aliquot of the ether-PE extract corresponding to 50 to 150 mg of total tocopherol is filtered through a 5 cm high column of alumina (ii). Total tocopherol of the column is washed out with 80 ml of eluting mixture No. 1.

Determination of α-Tocopherol

(a) ***Chromatography*** The filtrate obtained from (D) or an aliquot containing 75 ± 25 μg of total tocopherol is evaporated under vacuum, the residue is taken up in 20 ml of pure and distilled cyclohexane. The chromatographic tube is filled up to a height of about 12 cm by pouring slurry of alumina in cyclohexane of same grade. More cyclohexane is added until it covers the alumina for about 1 cm.

The cyclohexane solution of the sample is adsorbed on the alumina and washed with 60 ml of same quality of cyclohexane. This collecting flask is replaced by a clean one and the fraction containing α- and β-tocopherols is eluted with 130 ml of elution mixture No. 2. The other tocopherols remain adsorbed on the alumina.

(b) ***Colour reaction and measurement*** The eluate is evaporated under reduced pressure or in a stream of nitrogen, the residue is taken up in 8 ml of (i) (c). Addition of 1 ml each of α, α-dipyridyl (iv) and $FeCl_3$ solution (iii) is carried out, the absorption is measured exactly as done earlier under (A) (Procedure) (i).

Calculation

The calibration curve obtained by measuring the prescribed amounts of α-tocopherol is linear but does not pass through the origin, therefore, the Vitamin E content cannot be calculated directly from extinction readings found for the normal and the recovery tests.

μg of Vitamin E/100 g of sample = 100CB/A-B, where,

A = μg of Vitamin E/g of sample found in the recovery test by means of the calibration curve.

B = μg of Vitamin E/g of sample found in the normal test by means of the calibration curve, and

C = amount in μg of Vitamin E added in the recovery test/g of sample.

Accuracy of the method is ± 5%. Sensitivity is 10 μg Vit. E/g.

Note: The recovery test may be omitted when products of the same type have to be -assayed frequently, provided that the loss of Vitamin E during the whole assay is previously determined and does not exceed 7%.

Determination of total tocopherol It is determined by continuing the chromatography after elution of the α-and β-fractions. The collecting flask is changed, the other tocopherols are eluted with 80 ml of elution mixture no. 2. This eluate, which contains β, μ, δ, ε and n-tocopherols, is evaporated in vacuum and the residue is taken up in 8 ml ethanol (i) (c), and treated with ferric dipyridyl as described under (A) earlier. However, measurement of absorption is carried out 10 minutes after addition of the $FeCl_3$ solution because the tocopherols mentioned above react more slowly than α-tocopherol with ferric dipyridyl. The content of these tocopherols is calculated by reference to the calibration curve obtained with α-tocopherol.

Calculation

Total tocopherol content = Above result + this result, i.e.

α– and β tocopherol + β–, γ–, ε–, ε– and n-tocopherols.

ESTIMATION OF BIOTIN

The *lactobacillus arabinosus* has a specific requirement for biotin for its growth. The growth response on a defined medium, complete in all respects, except the biotin under test, is proportional to the concentration of the biotin added to the medium up to a certain stage. The acidity or turbidity produced by the organism is measured to determine the extent of growth and thereby the amount of biotin.

Reagents and Media

(i) ***Acid-hydrolysed casein*** Confirming to IS: 7203-1974, i.e. casein hydrolysate (acid digested), microbiological grade.

(ii) ***AGU solution*** It is prepared by heating 100 mg each of adenine sulphate, guanine HCl and uracil in a 250 ml Erlenmeyer flask containing about 75 ml H_2O + 2 ml conc. HCl. The resulting solution is cooled. In case a precipitate gets formed, it is brought into solution by repeatedly adding a few drops of HCl, heating and cooling. The cooled solution is transferred to a 100 ml flask and the volume made up with water.

(in) ***CT solution*** 4.0 g of L. cystine and 1.0 g of L-tryptophan suspension in 750 ± 50 ml water is heated to 75 ± 5°C, 20% HCl added dropwise with stirring until the solids are dissolved. Approximately 12 ml of 20% HCl is usually needed. Contents are cooled, volume made up to 1 litre with water.

(iv) ***Vitamin solution*** 20 mg riboflavin + 10 mg thiamine HCl + 10 mg of p-aminobenzoic acid + 40 mg pyridoxine HCl, all are transferred to a litre flask, the volume is made up with 0.02 N acetic acid.

(v) ***Pantothenic acid stock solution*** 54.4 mg of Ca-pantothenate diluted in 50% ethanol in water in a 500 ml volumetric flask and kept under refrigeration.

1 ml = 100 μg of pantothenic acid.

(vi) ***Niacin stock solution*** 50 mg niacin dissolved in 500 ml of ethanol: H_2O :: 1:1.

(vii) ***Biotin stock solution*** 25 mg of anhydrous *d-biotin* (free acid form) in ethanol : H_2O :: 1:1.

(viii) ***Salt solution** A 25 g* each of dibasic KH_2PO_4 + monobasic K_2HPO_4 solution in 500 ml of water. It is stored under toluene.

(ix) ***Salt solution B*** 10.0 g of $MgSO_4$ and 500 mg each of NaCl, $FeSO_4.7H_2O$ and $MnSO_4.4H_2O$ are dissolved in water, 5 drops of conc. HCl added and stored under toluene. HCl omitted when (i) is used.

(x) ***Basal medium for biotin assay*** For 100 assay tubes, the following ingredients are mixed in a 600 ml beaker.

50 ml each of (i) and (iii) + 10 ml each of (ii), (iv), viii and (ix) + 1 ml each of (v), (vi), + 10 g each of glucose and sodium acetate, both anhydrous. In practice, the ingredients are mixed thoroughly in 250 ml water, pH adjusted to 6.8 with 40% NaOH with the help of a pH meter. Contents are transferred to a 500 ml volumetric flask, the volume made up with water.

(xi) ***Either of the following two media can be used for stock culture.***

(a) 5.0 g yeast extract conforming to IS: 7004-1973 is dissolved in 200 ml H_2O, 1.0 g each of anhydrous glucose and sodium acetate + 3 g agar (IS: 6850-1973) are added, the mixture is treated on a steam-bath with occasional stirring until the ingredients get dissolved. The hot solution is filtered through cotton/cloth; 10 ml is distributed to each of 20 culture tubes. The tubes are plugged with cotton and autoclaved for 15 minutes at 120°C.

(b) The following ingredients are dissolved in 200 ml water, pH adjusted to 6.9, the volume made up to 250 ml with water.

Peptone (IS: 6853-1973) 5.0 g + yeast extract (as above) 1.0 g + anhydrous glucose and sodium acetate 10 g each + 2.5 ml each of (viii) and (ix).

(c) 7.5 g agar of (a) above grade is dissolved in 250 ml water by heating and mixed with (b) above. This hot solution (10 ml) is taken in each of 20 test tubes, plugged with cotton, autoclaved like before. After cooling to room temperature these are stored at 3 ± 1°C for future use for maintaining stock cultures.

(xii) ***Inoculum broth*** Its preparation is similar to culture medium (xi) above except that instead of agar solution, more water is used (after adjusting pH to 6.8) to make up the volume to 500 ml. 10 ml of this solution is taken into test tubes, plugged with cotton, autoclaved at 120°C for 15 minutes, cooled to room temperature and stored at 3 ± 1°C.

(xiii) ***Isotonic salt solution*** 900 mg of NaCl solution in 100 ml of water. 10 ml of this solution in each test tube is autoclaved as described at (xi) above.

(xiv) ***Bromothymol blue (BTB) indicator solution*** 100 mg BTB is dissolved in 1.6 ml of 0.1 N NaOH, It is triturated until the powder disappears, then diluted to 250 ml.

Alternatively, the dye may also be dissolved in 95% ethanol and diluted with 1.6 ml of 0.1 N NaOH followed by volume make up with water.

(xv) ***H_2SO_4 (6N)*** 9 litres of 6N H_2SO_4 is prepared by adding 1,500 ml of cone H_2SO_4 slowly and with stirring into 7 litres of water. The volume to 9 litres is made up with water only after cooling the solution.

(xvi) ***NaOH (20%)*** 40 g pellets + 160 ml water, contents stirred till complete dissolution.

(xvii) ***Working biotin standard*** 5 ml of (vii) is diluted to 250 ml with 50% ethanol. 5 ml of this solution is further diluted to 500 ml with 50% ethanol. Working standard is prepared afresh by diluting 5 ml of further diluted solution to 250 ml with water.

1 ml = 0.2 ng of biotin.

Procedure

(a) ***Preparation and maintenance of stock culture of Lactobacillus arabinosus (La)*** Pure culture of (La) 17-5 (ATCC 8014) is inoculated to the sterile agar culture tube prepared as described at (xi) (c). The tube is inoculated for 20 ± 4 hours at 37°C and stored at 4°C. The culture is transferred into new agar tube every fortnight.

(b) ***Preparation of inoculum*** One day before use (xii) tube is inoculated with culture from (xi) and incubated for·17 ± 1 hours at 37°C. A centrifuge tube plugged with cotton, all glass syringe and a 0.8 mm needle, some saline (.09%) is used in a conical flask. The saline is sterilized by autoclaving at 120°C for 15 minutes; syringe, needle, and centrifuge tube by heating in a hot air oven at 160°C for one hour. Cells from the inoculum tube are transferred to the centrifuge tube and centrifuged.

The supernatant is decanted, the cells are again suspended in sterile saline and recentrifuged. This process is repeated 2-3 times. The washed cells are again suspended in the saline, these resuspended cells are taken in the sterile syringe, the assay tubes are inoculated with one drop each of the inoculum whose optical density should be 0.11 ± 0.01.

(c) ***Preparation of sample*** Enough sample material and containing approximately 200 ng of biotin is pipetted/weighed as the situation warrants into a 50 ml conical flask. 25 ml of (xv) is added, contents thoroughly mixed, autoclaved at 120°C for 2 hours, cooled and finally transferred to 100 ml volumetric flask, volume made up and filtered if needed.

The degree of biotin homogeneity decides the quantity of material for extraction. Sample less than 2 g are avoided. For easy assay of biotin, high potency samples (like liver) are diluted.

A 10 ml aliquot is diluted to 65 ± 5 ml with H_2O in a 100 ml beaker, neutralized with (xvi) to pH 6.8, either with a pH meter or BTB indicator. The volume is made up to 100 ml in a volumetric flask.

Note: If the quantity of biotin in the original samples differs widely from 200 ng, it becomes imperative to dilute accordingly to get the final solution of 0.2 ng/ml content. The use of above two steps of dilution and procedure, rather than original dilution to 1 litre permits changes in the final dilution avoiding rehydrolyzation of the sample in the event of potency being too less than expected.

(d) ***Preparation of tubes for standard curve*** 0.0, 0.5, 1.0, 1.5, 2.0, 2.5,3.0,4.0, and 5.0 ml volumes of (xvii) are added to test tubes in duplicate so also 5 ml each of water and (x) to each tube. The tubes are covered with cotton and brown paper and sterilized at 115°C for 10 minutes.

(e) ***Preparation of assay tubes*** To duplicate tubes is added 1.0, 2.0 and 3.0 ml aliquots of the test solution. The use of these different levels permits evaluation of the validity of the assay over a range of biotin concentration within the limits of the standard curve. Where the approximate potency of the sample is not known, it is suggested that test solution over a greater range of concentration is added, i.e. 0.5, 1.0, 2.0, 3.0, 4.0 and 5.0 ml if the batch is of 6 tubes.

(*f*) Water is added to make up the volume to 5.0,1 ml in each tube. 5.0 ml of (x) is added to each tube and proceeded like in (d).

(*g*) ***Inoculation and incubation*** Each tube is inoculated aseptically with one drop of inoculum except the blank, incubated at 37°C for 19 ± 1 hours.

A. Titrimetric method Contents of each tube is transferred to 125 ml conical flask. The tubes are rinsed with about 10 ml H_2O, the rinsings added to the flask, about 0.2 ml of 0.1% BTB mixed, contents titrated against 0.1N NaOH, to a green colour, approximate pH 6.8. One such flask is enough for colour comparison for about 15 titrations.

Calculation

A standard curve is drawn by plotting ml of 0.1 N NaOH required to titrate contents in the standard tubes against ng of biotin in the titrated tube. The biotin content in the unknown series is determined by interpolation of the titre values on the standard curve. Values more than 1.00 ng or less than 0.10 ng biotin/tube are discarded. The biotin content of each set is determined in duplicate and of the test material from the average of the values for each of the solution obtained from not less than 3 sets of these tubes which do not vary by more than 10% from the average, using the following formula.

$$\text{ng of biotin/g of sample} = \frac{\text{Average ng} \times \text{volume} \times \text{Dilution factor}}{\text{Mass of sample}}$$

B. Turbidimetric method At the end of incubation period (procedure) (g), the assay tubes are removed and steamed for 5 minutes, cooled and turbidimetrically read in a colourimeter at 660 nm using the non-inoculated blank tube for zero setting of the instrument.

Calculation

A standard curve for the assay is plotted, the optical density/turbidity reading on the X-axis against concentration of the biotin on the Y-axis. The biotin content of the tubes in the unknown series is determined by interpolation of the colourimeter readings on the standard curve. Average for 1 ml of the test solution is calculated from values obtained from a minimum of 3 sets of tubes. These values must not vary more than 10% on the average.

ng of biotin/g of sample

$$= \frac{\text{Average ng/ml} \times \text{Diluting factor}}{\text{Mass of the sample}}$$

Since it is not possible to duplicate exactly from time to time conditions of autoclaving, temperature for incubation, etc., influencing the curve readings, it is imperative that a curve is constructed each time an assay is undertaken.

BETALAINS

Extraction

Separation and purification of betalains from crude plant extracts is usually necessary if qualitative and/or quantitative analyses are to be carried out.

Preliminary separation of betalains in a variety of crude aqueous plant extracts have been accomplished by non-ionic absorption on strongly acid resins like Dowex 50 W, followed by polyamide column chromatography, using increasing concentrations of methanol in aqueous citric acid, as the developing solvent since 1964. Citric acid is removed from resolved betalains by resin treatment. Sequential chromatography on series of Sephadex ion exchanges has also provided good separation of betalains. Adams *et. al.* have reported that gel filtration column chromatography or Sephadex or poly amide (Bio-Gel P-6) gels could be used to rapidly and efficiently separate the betalains from beef juice. Numerous other chromatographic and electrophoretic procedure, have been used over the years to separate/purify betalains from a variety of sources but are time consuming. Since their application for analysis of the betalains, HPLC methods have become widely adopted as a fast and efficient means of betalain separation and analysis.

Qualitative Analysis

Structural characterisation and identification of betalains generally involves direct comparison of spectroscopic, chromatographic and electrophoretic properties with authentic standards, before and after controlled hydrolysis. Since the betalains differ from anthocyanins in being more sensitive to acid hydrolysis, in their colours with changes in pH, and in their chromatographic and electrophoretic properties, they are easily differentiated by simple colour tests. Since the occurrence of these two pigment types is mutually exclusive, colour tests can usually be performed using crude plant extracts.

Betacyanins and betaxanthins possess similar chromatographic and electrophoretic properties, thus, similar techniques are used for their isolation/purification. However, polyamide chromatography is only moderately successful for the separation of betaxanthins. Preparative paper electrophoresis and chromatography using alternate supports are commonly used for separation of individual betaxanthins from co-occurring betacyanins and other plant constituents. HPLC is also becoming an invaluable means to separate and analyse the betalains, the number of betaxanthin structures was found to be considerably greater using HPLC than was known from more traditional separation techniques. The spectral properties and RP-ion pair HPLC retention characteristics of a number of natural and semisynthetic betaxanthins were recently reported[9].

Tentative identification of betalains can be deduced from their chromatographic or electrophoretic behaviour. For example, on polyamide or cellulose-based resins:

(i) The retention of betacyanins decreases with increased glucosyl substitution.

(ii) The retention of 6-glycosides exceeds that of 5-glyco-sides.

(iii) Iso-derivatives are retained slightly longer than their corresponding C-15 epimers.

(iv) Acyl groups, aromatic more than aliphatic, increase the retention of pigments.

Each pigment possesses characteristic electrophoretic mobility between 0.30 and 1.78 relative to betanin, usually determined at both pH 4.5 and 2.4.

Corroborative data may be provided by analysis of absorption spectra. All betacyanins exhibit visible maxima between 534 and 552 nm, while all betaxanthins have absorption maxima in the range of 474 to 486 nm. Acylated betalains generally exhibit a second absorption maximum in the UV region of 260 to 320 nm, where the absorption of non acylated pigments is weak. As with the anthocyanins the number of acyl residues in betalains can be estimated from the ratio of absorption at the visible maximum to that at UV maximum.

The current sensitivity of NMR spectrometers permits complete structural characterisation of betalain pigments, even when sample size is small. Alard *et. al.* and Strack *et.. al.* have obtained corroborative molecular weight data by FAB MS and/or GC/MS. The former technique has the advantage that the betalain pigments do not require derivatisation before hand.[10,11]

The IR spectra of betalains have been found to provide little useful information.

Quantitative Analysis

In the absence of interfering substances like fresh extracts or juice, purified samples, the quantitation of betalains can be accomplished using spectrophotometry, whereby absorbance at the visible maximum is expressed in terms of concentration by use of appropriate absorptions as reported by several workers.

T. Nilsson developed a spectrophotometric method that directly determines betacyanin and betaxanthin pigments in beets without their initial separation. This method is based on the observation that while vulgaxanthin-I absorbs only at 476-478 nm, betanin exhibits a maximum at 535-540 nm but also absorbs at the visible maximum of vulgaxanthin-I. Calculation of ratio of absorbance at 538 nm to that at 477 nm as a function of concentration, i.e. dilution leads to determination of the measurable concentration of vulgaxanthin I. Absorbance of the sample at 538 nm is due only to betanin, permitting its direct determination. Results are expressed in terms of total betacyanin and total betaxanthin concentration since minor constituent betalains also contribute to measured absorbances.

If interfering substances like betalain pigments are present in the sample, they must first be purified. The procedures generally used, usually paper/column chromatography or electrophoresis, not only purify but also separate pigments, thereby allowing for quantitation of total and individual betalains simultaneously. Due to its speed and high resolving power, and the fact that preliminary purification is not always necessary, HPLC is the method of choice for quantitative analysis of individual and total betalains.

Determination Iron

Reagents

Standard iron stock solution

(a) 100 mg of pure iron wire is dissolved in 5 ml of dilute HCl (1:1) (v/v) by heating in a small flask. 250 mgm of $KClO_3$ is added, contents boiled until free from chlorine odour and diluted to 1 litre in a graduated flask with water.

1 ml = 100 μg of iron.

(b) Working standard when necessary, it is prepared fresh by diluting 10 ml of (a) to 50 ml with water in a graduated flask.

1 ml = 20 μg of iron.

Procedure

7.5 ± 2.5 g of the sample accurately weighed is incinerated in a platinum basin at a temperature not exceeding dull redness until the ash is white or nearly so. After cooling, 5 ml of dil. HCl (1:1) (v/v) is added followed by 1 ml of dil. HNO_3 (1:1) (v/v), contents heated on a water bath and stirred well to dissolve the ash completely, cooled, volume made up to 100 ml in a graduated flask with water and mixed. This is filtered through Whatman filter paper No. 40 or its equivalent. This is solution A.

10 to 25 ml of A, corresponding roughly to 0.1 mg of iron, is transferred to a Nessler cylinder, 1 ml of conc. HCl, sp. gravity 1.16 is added then diluted to 100 ml mark with water. 5 ml of NH_4CNS, 50%, w/v is added and mixed. If the painting colour is too intense, a small quantity of the solution should be used. The colour is matched with that of known amounts of the standard solution treated similarly. For this purpose known amounts of the standard iron solution are mixed with 1 ml of conc. HCl, diluted to the 100 ml mark in a Nessler cylinder and then mixed with 5 ml of NH_4CNS solution.

Calculation

(a) From the known amounts of the standard iron solution the iron (Fe) content of the sample is calculated as,

$$\text{Iron (Fe) mg/100g} = \frac{10000 \times \text{concentration of iron (Fe) in aliquot}}{VW}$$, where,

V = volume in ml of the solution A taken for the test, and
W = weight in g of the material taken for the test.

(b) Alternatively, a photoelectric colourimeter may be used to construct a calibration graph using known amounts of the standard iron solution against the optical density readings observed with each concentration at 480 mu. The iron content of the test solution is found from graph after observing the optical density reading. The calculations remain the same as under (a) abcve.

Determination of Calcium

25 to 50 ml solution A are transferred to a 250 ml beaker, diluted to approximately 100 ml with water and two drops of methyl red solution (1 g in 200 ml ethanol): NH_4OH (1:1; v/v) is added dropwise to get solution pH 5.6 as shown by intermediate brownish-orange colour. In the event of overstepping, HCl, 1:3; v/v is added with a dropper to orange point. Two more drops of this HCl are added when pink colour, pH 2.5 to 3.0 and not orange appears. This solution is diluted to 150 ml, brought to boiling and with constant stirring is slowly added 10 ml hot saturated aqueous ammonium oxalate. If the red colour changes to orange or yellow, HCl,1:3; v/v is added dropwise until colour changes to pink. The precipitate is kept aside overnight for settling. The supernatant liquid is filtered through quantitative filter paper/Gooch/fritted glass filter, the beaker is washed, then precipitated thoroughly with NH_4OH, 1:5 until free from oxalate. The filter paper/crucible with precipitate is placed in the original beaker, 130ml of 1.5N H_2SO_4 is added, contents heated to 70°C or above and titrated with 0.05 N $KMnO_4$ to first slight pink colour. Blank correction is applied.

To the hot filtrate is added 10 ml of saturated solution of ammonium oxalate, boiled for a minute and well stirred. The precipitate is let to form and settle. The supernatant liquid is decanted through 11 cm Whatman No.40 filter paper or its equivalent. The precipitate is once washed by decantation with warm water and then transferred to the filter paper. The beaker is washed thoroughly with warm water several times passing the solution through the filter paper. The filter paper and the precipitate is washed with hot water until the filtrate is free from ammonium oxalate, as tested by the standard $KMnO_4$ solution. The beaker in which calcium precipitation was done is placed under the funnel, the apex of the filter paper is pierced with a pointed thin glass rod, the precipitate is washed into the beaker with hot water. 25 ml of the hot dilute H_2SO_4, 1.5N is added dropwise into the filter, taking care that the acid comes into contact with every part of the filter paper and that some of it is poured behind the doublefold of the filter paper, in case any of the calcium oxalate is lodged therein. The filter paper is washed thoroughly with hot water. Calcium oxalate solution is heated to about 90°C and titrated with the standard $KMnO_4$ solution to a faint pink colour.

Calculation

Calcium percentage = 10 X/VW, where,

X = volume in ml of 0.05 N$KMnO_4$(1ml of 0.05N $KMnO_4$ = 0.001 g of calcium),

V = volume in ml of the solution A taken for the test (above) and

W = weight in g of the material taken for the test (iron procedure).

References

1. Pfander H and R Reisen, *Carotenoids, Vol.1 A-Isolation and Analysis,* pp 81-108, 1995.
2. Nelis HJCF and De Leenheer AP, *Analytical Chemistry,* pp 270-275,1983.
3. Britton G and TW Goodwin, *Plant Pigments,* Academic Press, London, pp 61-132, 1988.
4. Britton G, *Natural Food Colourants* (Carotenoids), 2nd Edn, edited by Hendry GAP and Houghton JD, Blackie Academic & Professional London, pp 222-223,1996.
5. Davies BH, *Chemistry and Biochemistry of Plant Pigments,* 2nd Edn., Goodwin TW (Ed). Vol. 2, Academic Press London, pp 38-165,1976.
6. Britton G, *Methods Enzymol,* pp 111, 113-149, 1985.
7. Britton G, *Carotenoids, Vol.1 B: Spectroscopy,* Brikhauser, Basel, Boston, pp 13-62, 1995.

8. De Ritter E, *Carotenoids as Colourants and Vitamin A Precursors,* 'Academic Press, New York, pp 883-923, 1981.
9. Trezzini GF and Zryd JP, *Phytochem,* pp 1901-1903, 1991.
10. Alard D, Wray V, Grotjahn L, Reznik H and Strack D, *Phytochem,* pp 2383-2385,1985.
11. Strack D, Schmidt D, Reznik H, Boland W, Grotjahn L and Wray V, *Phytochem,* pp 2285-2287,1987.

Chapter 10

Miscellaneous Detections

QUININE

It is also known as Conquinine; pitayine; β-quinine. It is the most important alkaloid of Cinchona bark, the bark of *Cinchona officinalis (L); C. ledgeriana (Moens)* which contains about 8% quinine, other barks 1-4%.

Quinine is triboluminescent, orthorhombic needles from absolute alcohol, mt. pt. 177°C (some decompose). Sublimes in high vacuum at 170-180°C. pH of saturated aqueous solution 8.8. The blue fluorescence is especially strong in dil. H_2SO_4. 1 g dissolves in 1900 ml water, 760 ml boiling water, 0.8 ml alcohol, 80 ml benzene (in 18 ml benzene at 50°C), in 1.2 ml $CHCl_3$; 250 ml dry ether, 20 ml glycerol, 1900 ml of 10% ammonia water. Almost insoluble in petroleum ether.

It is bitter, stomachic, analgesic, antipyretic in veterinary medicine. It is also an anti malarial drug. It is permitted in soft drinks and tonic waters. It was perhaps Coca Cola drinks which used quinine in its soft drinks for imparting a better taste as remembrance element to make the taste of the drink linger in the mouth. At that time quinine phosphate and its recent derivative diphosphate were not available so the manufacturers had to be content with quinine itself. Now a days quinine diphosphate is used in quantities not more than 16.5 mg/300 ml though there is no statutory impingement in the matter.

Determination Quinine is determined usually by fluorimetry, though several spectrophotometric methods are also available. For this purpose either a fluorimeter with 365 nm excitation filter and 415 nm emission filter or spectrofluorometer is required.

Fluorimetric method

Reagents

(i) ***H_2SO_4 (0.05M)*** 2.78 ml is diluted to 1 litre with water.

(ii) ***Standard solutions of quinine sulphate.***

(a) ***Stock solution*** 50 mg of quinine or 60.4 mg of quinine sulphate dihydrate is transferred into 500 ml volumetric flask, 25 ml of 1M H_2SO_4 added, contents diluted to volume with water.

(b) ***Working standard*** Dilute (a) above with 0.05M H_2SO_4.

Preparation of sample Carbonation is removed by stirring and 5 g of the sample is accurately weighed, diluted to 250 ml in a volumetric flask with 0.05M H_2SO_4. 5 ml of this homogenised solution is further diluted to 25 ml using same 0.05M H_2SO_4.

Procedure

Using 365 nm excitation filter and 415 nm emission filter, the fluorimeter is adjusted to zero with 0.05M H_2SO_4. The emission intensity is measured against concentration, µg. The fluorescence of the sample solution is read, the amount of quinine present in the sample is calculated from the calibration curve.

Spectrophotometric method Quinine is extracted from soft drinks using $CHCl_3$ after making it alkaline with NaOH. It is reextracted from the organic layer with IN H_2SO_4. An aliquot of acid extract is adjusted to pH 1.0, alizarin brilliant violet R (ABVR) solution is added and the complex is extracted into $CHCl_3$. The absorbance of $CHCl_3$ layer is measured at 578 nm. Amount of quinine present in the sample is computed from the calibration graph.

Reagents

(i) Buffer (pHl): 97.0 ml of 1.0 N HCl + 50 ml of 1.0 M KCl are mixed and diluted to 1 litre with water.

(ii) ABVR solution (1 x 10^{-3}M) in water.

(iii) Quinine standard solution (1 mg/ml in 0.1 N H_2SO_4).

(iv) Working standard (40 ug/ml.)

Procedure

(a) In a 100 ml separating funnel (SF), 30 ml (i) + 5 ml of dye solution + aliquot of (iv) ranging from 10-200 μg. The volume is adjusted to 50 ml with water. Extraction with 10 ml of $CHCl_3$ is done, the absorbance of $CHCl_3$ layer is read at 578 nm and from these a calibration graph is plotted.

(b) 50 g of the soft drink sample is accurately weighed and transferred to a 250 ml SF. 10 ml of 10% NaOH is added, mixed and three extractions each with 20 ml of $CHCl_3$ are done. The three $CHCl_3$ phases are combined and shaken with 3 x 30 ml of IN H_2SO_4 in another SF, the acid solutions are taken in a 100 ml volumetric flask, the volume made up with the acid. An aliquot of this solution is treated as at (a). 1 to 5ml of the soft drink sample is processed directly as at (a).

The amount of quinine present in the sample is calculated from the absorbance of sample using standard curve obtained at (a).

METHYL IMIDAZOLE IN CARAMEL

It is a contaminant in caramels produced by NH_3 and $(NH_4)_2\ SO_3$ processes. B.I.S. has specified 300 ppm (max) in such caramels. It is determined by adding caramel colour to a basic celite column and eluted with a mixture of $CHCl_3$ and alcohol. The eluant is extracted with dil H_2SO_4 and the aqueous extract is made to a known volume. Taking aliquot of this extract, colour is developed with diazotised sulphanilic acid in alkaline medium. The amount of 4-methyl imidazole is calculated from the curve.

Apparatus A spectrophotometer and a chromatographic column measuring 25.0 x 2.5 cm, with a stop cock along with usual glass ware is needed.

Reagents

(i) *(a)* ***Chromogenic reagent*** 1 g sulphanilic acid dissolved in 9 ml of conc. HCl is made upto a known volume and cooled to 4°C.

(b) 1 mg of $NaNO_2$ in 100 ml H_2O, cooled to 4°C, 25 ml each of (a) and (b) are mixed afresh before use.

(ii) *(a)* ***Stock solution of 4-CH_3 imidazole (4Mi)*** 100 mg of 4 Mi is dissolved in 100 ml of 0.1N H_2SO_4 and stored in a refrigerator.

(b) ***Standard solution*** 5 ml of (ii) (a) diluted to 100 ml so that the resultant concentration is 50 μg/ml.

(c) ***Working standard solution*** 10 ml of (ii) (b) diluted to 100 ml; 5 μg/ml

(iii) celite 545.

(iv) Elution mixture $CHCl_3$: alcohol : : 4 : 1.

Procedure

3 g of celite 545 + 2 ml of 2N NaOH are mixed well. A fine glass wool plug is placed at the base of the chromatographic column followed by 5 g of celite-NaOH mix prepared earlier, tagging this material mass firmly and uniformly. 10 g of the caramel sample in 100 ml beaker + 6 g of 20% Na_2CO_3 solution are mixed and packed on the column. The beaker is washed with 2 g of celite, the washing also added to the column, which is sealed at the top with another glass wool plug and eluted with (iv) till 150 ml of eluate is collected into a 250 ml beaker. The eluate is transferred to a 250 ml separating funnel (SF) and extracted with 3 x 20 ml portions of 0.05N H_2SO_4. This extract should be strongly acidic (pH3) confirmed by test paper. The combined aqueous extract is concentrated by employing fresh evaporator at less than 50°C. The residue's volume is made up to a known volume.

Preparation of standard curve

0.0, 1.0, 2.0, 3.0 and 5.0 ml of (ii) (c) is taken in a series of 25 ml volumetric flasks, 1.0 ml of (i) (a) + 2.0 ml of Na_2CO_3, 5% solution in each flask, the volume is made up. Absorbance at 505 nm is read, the standard graph is plotted.

Colour of 2.0 to 5.0 ml solution of the aqueous extract is developed, its absorbance read, whence the quantity of 4 Mi present in the sample is calculated.

DETECTION OF BROMINATED VEGETABLE - OILS (BVO) IN SOFT DRINKS

BVO in the soft drinks is extracted with diethylene. The concentrated ethereal solution is treated with a small quantity of zinc dust to convert the organic bromide to inorganic form and treated subsequently with PbO_2 to liberate bromine detectable with fluorescein treated filter paper which turns pink due to formation of eosine.

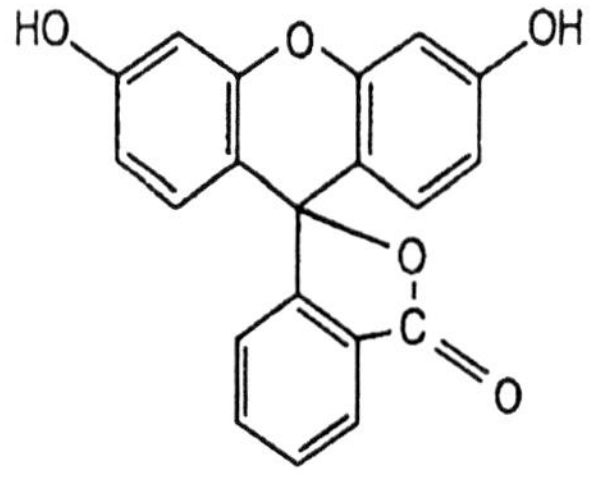

HO, O, OH, O, C, O + 4 Br → NaO, Br, Br, O, O, Br, Br, COONa

Fluorescein

Figure 10.1

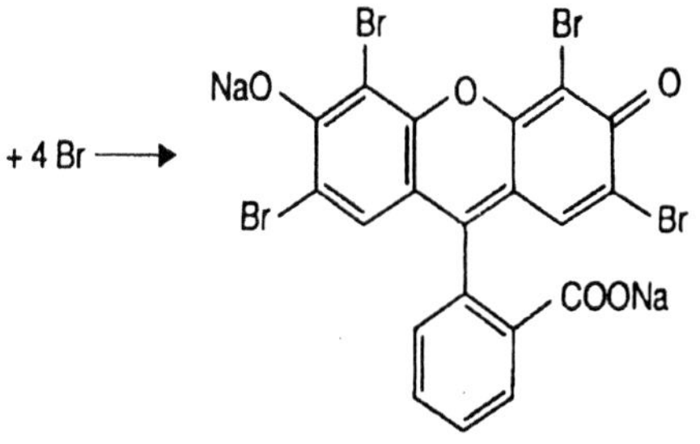

Eosine

Figure 10.2

The following test can detect as low as 1 ppm BVO.

Procedure

Fluorescein treated filter strips are prepared by preparating its 0.1% solution in ethanol : H_2O :: 1:1, the solution is made alkaline by adding 1 ml of 0.1 N NaOH solution per 100 ml. In the filtered solution, Whatman No. 1 filter paper strips, 2 x 10 cm are soaked for 10 minutes. The strips are air dried before use. If needed regularly then the dried strips are stored in an air tight brown bottle. Pale or discoloured strips are not suitable.

200 ml or a suitable aliquot of the soft drink, containing about 2 mg BVO is transferred into a separating funnel (SF). Extractions with 2 x 75 ml portions of diethyl ether are washed with 25 ml water and combined, dried by filtration through anhydrous Na_2SO_4. The ethereal extract is concentrated to approximately 5 ml in a conical flask, 25-30 mg zinc dust and 5 ml of dil acetic acid added to the extract, heated for 5 minutes over a steam bath. The contents are cooled, filtered through Whatman No. 1 filter paper into another glass stoppered 100 ml conical flask. The filter paper is washed with 2-3 ml of dil. acetic acid rinsings of the original flask. 25 to 30 mg PbO_2 is added to this flask and the above paper strip from the earlier described air tight bottle is hung immediately inside this flask along with the stopper, the strip must not touch the liquid in the flask. The flask along with contents is heated over a steam bath; change in strip colour is observed. The yellow strip turns to pink within 3 minutes if BVO is present in the sample.

SODIUM DIACTATE (SDA) CH_3COONa | CH_3COOH

SDA prevents mold growth in malt syrup at 0.5% level, in cheese spreads at 0.15 ± 0.05% level. It is very effective in baking industry by virtue of its inhibitory power against bread mold and rope, being practically inert to baker's yeast. At 3250 ± 750 ppm levels it is used in bread and cakes. In India it is allowed in bread upto 4000 ppm and in flour for baked food upto 2500 ppm.

SDA or Dykon may be understood as a 'bound' compound of sodium acetate and acetic acid. It is a white powder, decomposes above 150°C, soluble in water liberating 42.25% available acetic acid.

Glab[1] has described it as sequestrant.

NISIN

Nisin C_{143} H_{250} N_{12} $O_{,27}$ S_{7}, mol. wt. 3354.25 is a polypeptide antibiotic structurally similar to subtilin minus tryptophan. Structure contains 34 amino acid residues, 8 of which are rarely found in nature, including lanthionine (two alanines bounded to sulphur at the p-carbons) and p-CH_5-lanthionine. It crystallises from ethanol, is soluble in dilute acids and is stable to boiling in acid solution.

Nisin up to 12.5 ppm has been allowed in *paneer*/channa/cheese-sliced/cut/shredded/processed and spread and in canned rasgulla 5 ppm. In prepared coconut water it has been allowed up to 5000 I.U.

It is a polypeptide type antibiotic produced by *Streptococcus lactis*. It is mostly active against gram-positive bacteria, many lactic acid bacteria, strepto cocci, bacilli, clostridia and other anaerobic spore forming microorganisms responsible for spoilage in canned food and dairy products especially processed cheese. Yeast and mold are not inhibited by nisin, in fact many of them tend to decompose it readily. Proteolytic enzymes like trypsin, pancreatin, digestive enzymes and salivary enzymes but not rennet, hydrolyse and inactivate nisin. The action of nisin is directed against the cytoplasm membrane and consequently nisin is more active against spores than vegetative cells. Addition of nisin in canned foods is reported to reduce thermal processing time for sterilization.

Consequently major use of nisin all over the world is in the preservation of processed cheese, where primary action of nisin is directed against spores forming microbes mainly clostridia and butyric acid species. Its use as an auxiliary sterilant in canned fruits and vegetables is doubtful though used occasionally.

Nisin can be assayed after extraction from processed cheese with hydrochloric acid by a plate diffusion technique using *Micrococcus flavusas* test organism, BS 4020: 1974. For sorting purposes Tramer and Fowler, 1964 have described a reverse-phase disc assay technique based on B. *stearothermophilus* spores with overnight incubation at 55°C. Stumbo *et al.* 1964 have described a sensitive method employing *heat-damaged* spores of *B thermophilus* for determining nisin in numerous canned foods.

Reference

1. Glab, *Food Industry,* 14, No.2, pp 46, 1942.